FACHRS School Logbook

SCHOOL and COMMUNITY

Family and community history through the prism of school logbooks

FACHRS
FAMILY & COMMUNITY HISTORICAL RESEARCH SOCIETY

FACHRS Publications
Pilcot, Hants

Published by
FACHRS Publications
Family & Community Historical Research Society Limited
Pilcot House, Pilcot, Dogmersfield, Hook, Hants RG27 8SY

ISBN: 978-0-9548180-3-6

Front Cover Illustration:
Photograph of a recreated classroom in Stirchley Board School. The school was opened in 1881 and closed in 1973. The building was dismantled and moved to its present location at the Blists Hill Victorian Town, part of the Iron Gorge Museums Trust, in 1992.
Photograph © Brita Wood 2010

Back Cover Illustration:
Clifton on Dunsmore Village School, Warwickshire, built 1850.
Photograph © Anne Langley 2018

Cover design and typesetting:
Angela Blaydon

Set in Times New Roman 10pt
Printed on paper from sustainable sources

FSC
www.fsc.org
MIX
Paper from
responsible sources
FSC® C020438

Printed by Hobbs the Printers Ltd, Totton, Hants, SO40 3WX
www.hobbs.uk.com

CONTENTS

LIST OF ILLUSTRATIONS, TABLES AND FIGURES

FROM ANOTHER ANGLE: FAMILY AND COMMUNITY HISTORY THROUGH THE PRISM OF SCHOOL LOGBOOKS
Michael Drake

Introduction

The ostensible purpose of school logbooks was to provide a daily record, penned by the principal teacher of a school, of the education carried out by him or her and any other staff, and the factors affecting it. However, many school logbooks did far more than this, in part because many teachers did not follow the instruction printed in each logbook, nor heed the strictures of the reports of the school's inspector, a summary of whose annual report on the school was also included in each logbook (*see* Illustration 1).

Matters appearing in logbooks of relevance to the history of family and community were 'weather reports, meets of hounds, haymaking, harvesting and gleaning, planting and picking, hiring fairs, wooding, beating cover for various estates, and illegal employment by neighbouring farmers'. All these were instanced by a school inspector when commenting on the 'everlasting problem of the school attendance of rural children'.[1] And there was more, such as local fairs, wakes and epidemics.

Extracts from the Revised Code of Regulations announced in Parliament on the 13th February and 28th March, 1862, and confirmed by the Right Honourable the Lords of the Committee of the Privy Council on Education in the Minute of the 9th May, 1862.

CHAPTER II. PART I. SECTION I.

55. In every school receiving annual grants is to be kept, besides the ordinary registers of attendance,—
   (a) A diary or log-book.
   (b) A portfolio wherein may be laid all official letters, which should be numbered (1, 2, 3, &c.) in the order of their receipt.

Diary or Log Book of School.

56. The diary or log-book must be stoutly bound and contain not less than 500 ruled pages.
57. The principal teacher must daily make in the log-book the briefest entry which will suffice to specify either ordinary progress, or whatever other fact concerning the school or its teachers, such as the dates of withdrawals, commencements of duty, cautions, illness, &c., may require to be referred to at a future time, or may otherwise deserve to be recorded.
58. No reflections or opinions of a general character are to be entered in the log-book.
59. No entry once made in the log-book may be removed nor altered otherwise than by a subsequent entry.
60. The inspector will call for the log-book at his annual visit, and will report whether it appears to have been properly kept throughout the year.
61. The inspector will not write any report on the good or bad state of the school in the log-book at the time of his visit, but will enter therein with his own hand the full name and standing (*certificated teacher of the —— class, or pupil-teacher of the —— year, or assistant-teacher*) of each member of the school establishment. The inspector will not enter the names of pupil-teachers respecting whose admission the Committee of Council has not yet pronounced a decision.
62. The summary of the inspector's report when communicated by the Committee of Council to the managers must be copied into the log-book by the secretary of the latter, who must also enter the names and description of all teachers to be added to, or withdrawn from, those entered by the inspector, according to the decision of the Committee of Council upon the inspector's report. The secretary of the managers must sign this entry.
63. The inspector before making his entry of the school establishment in the following year will refer to his own entry made in the preceding year, and also to the entry which is required to be made by the secretary of the school pursuant to Article 62, and he will require to see entries in the log-book accounting for any subsequent change of the school establishment.

Illustration I. *Leigh National (Mixed) School Logbook 1862 (Surrey) instructions.*
Reproduced by kind permission of Surrey History Centre
© *Surrey History Centre*

Logbooks were introduced in 1862 into schools receiving government grants, by the somewhat notorious Robert Lowe.[2] They were intended to be 'stoutly bound' and 'not less than 500 ruled pages'. According to the 'Revised Code':

1. The principal teacher must daily make in the log-book the briefest entry which will suffice to specify either ordinary progress or whatever other fact concerning the school or its teachers, such as the dates of withdrawals, commencements of duty, cautions, illness etc. may require to be registered at a future time, or may otherwise be recorded.

2. No reflections or opinions of a general character are to be entered in the log-book.

3. No entry once made in the log-book may be removed nor altered, otherwise than by a subsequent entry.

So far as the school inspectors were concerned they were to call for the school's logbook at their annual visit and report whether or not it had been kept properly throughout the year. A summary of the inspector's annual report, when communicated by the Committee of Council to the school managers, was to be pasted into the logbook for that year by the secretary of the latter.[3]

Teething problems?

It could hardly be expected that the introduction of school logbooks would be trouble-free. Even before they appeared in schools it was argued that the instructions to the principal teachers were too detailed or not detailed enough.[4] Once in operation, school inspectors soon found plenty to criticise. One, whose job it was to inspect schools in what is now Cumbria remarked in his 1864 report that:

I have been much disappointed at the way in which the logbook had been kept during the past year. I am quite aware that in small country schools, especially where there are no pupil-teachers, it may sometimes prove difficult to find material for every day without falling into impertinent general observations [what on earth can he mean?] but I am sure that the log-book might, even in these circumstances, be much more intelligently kept than it is. A great deal depends on the animus with which a teacher sets to work. If he chooses to think the log-book unnecessary, useless or inquisitive, of course he will "not know what to enter." If, on the other hand, he regards it as an excellent aid towards systematising his work, and towards proving to the inspector, that he does so, and as a means of showing how much methodical thought and intelligence he bestows on all the details of his profession, he will find plenty to say. During the past year there must have been a want of some of these qualities in the teachers of this district, or else a considerable prejudice against the logbook. There is no other sufficient reason why logbooks should often have been so inadequately kept. There have, however, been some brilliant exceptions to this general rule; and I take with pleasure this opportunity of mentioning the names of the following good teachers who, from the first, kept their logbooks in such a style as to give me a high sense of their intelligence and efficiency.

Mr. J. Green, National School, Brough in Westmoreland.
Miss Ann Joy, Fawcett Girls' School, Carlisle.

> Of these Miss Joy's was decidedly the best I have seen throughout the year, and was in every way, like all her school-work, thoroughly satisfactory.[5]

This theme appeared often. For instance, a Scottish inspector complained that:

> Logbooks might be better kept and made more profitable. Too often have I found an entry or two wanting. Sometimes the entries for pages consist of "Lessons given according to the time-table. Usual progress." Or chronicle irrelevant incidents. One in Ross-shire is mainly devoted to the crops. Another, aiming at elevation and precision, gravely informs one that on a certain day the attendance was "decimated" by "one half".[6] A Welsh inspector, after praising those teachers who kept 'their logbooks well', noted that some 'had no sense of discretion as to what they shall enter'.[7]

Yet another inspector noted:

> On one or two occasions I have found it necessary to complain to your Lordships of the personal entries, characterized by great insolence, which have been made in their logbooks by certain schoolmasters, and in one case I forwarded a logbook to the council offices. Such a thing as this is easily prevented by managers looking over the logbooks carefully from time to time.[8]

The FACHRS logbook project inaugurated by its then chairman, Donald Dickson, produced transcripts covering twenty-six schools[9] using an Access database devised by him. Following this Introduction are four articles by members of FACHRS who took part in the project.

Of the transcripts, several types of entry are of interest to family and community historians, apart from the ones listed above. For instance, logbooks often mentioned the names of visitors to the school, although they did not always report why or what they did there. Some were from the upper classes. For instance, on 6 June 1873, the Albert Lane School in Saltaire, West Yorkshire was visited by 'Lord Lyttleton, Lord Frederick Cavendish, Sir Andrew Fairbairn and other distinguished gentlemen'.[10] These may have been the guests of Sir Titus Salt, industrialist and philanthropist, who had built the model village of Saltaire and who also visited the school on 27 February 1874. His son, Titus Salt Jnr., was the secretary of the school.[11] Lyttelton had addressed the House of Lords in a debate on the introduction of the Revised Code, complaining that 'the diary and logbook were not simply or fully explained'.[12] He committed suicide by throwing himself down a staircase. Cavendish was famously assassinated by a secret political society in Phoenix Park, Dublin on 6 May 1882.[13]

More commonly, schools were visited by clergyman. For instance, the Rev. HS Toms was a frequent visitor to the Bush Hill Park Board School for Girls in Enfield, Middlesex.[14] Toms was a Congregational Minister. Another clerical visitor to the same school was the Rev. Prebendary George Hewitt Hodson, vicar of Enfield, author of *Hodson of Hodson's Horse: twelve years of a soldier's life in India*, a classic text of the Raj. This was his brother WSR Hodson.

Of course, not uninteresting tittle-tattle of this sort can be elicited from a great many historical sources, but this is especially the case as regards school logbooks, because they often reflect the idiosyncratic character of those responsible for compiling them. However, as the transcripts of the logbooks and the articles below indicate; taken together with other sources they can help to provide important insights into aspects of community. The most significant of these are those the schools' inspector indicated in the Leicester newspapers in 1900 (see above).[15] How they can be used for family and community history can be seen from the articles below.

For instance, Frances Brooks, in her 'Window on community: the chronology of school holidays 1873-1921' shows how still, in the late nineteenth century, parents, employers, and the strength of local custom, played a significant part in determining the conduct of at least one local institution, the school, evidenced by the wide variation in both the timing and length of school holidays. As holidays are one of the matters covered almost universally by school logbooks, this is an issue that could be replicated widely.

Rob White, in his 'Jobs for the girls? The recruitment of Teachers in Winchcombe, Gloucestershire 1860-1910' takes up an important issue in nineteenth century life, the role of nepotism in the job market. Using the school logbooks as his core source, he makes a convincing case for the central role of nepotism in starting off the careers of teachers, namely through their appointment as monitors and pupil teachers by family members. Although the logbooks used appear to be particularly well-kept and detailed, so allowing him to draw quantitative conclusions, the role of qualitative evidence, in the form of family histories, cannot be under-estimated.

Ray Greenough in his 'It's not my fault that I don't turn up at school'; a school logbook perspective on pupil attendance between 1880 and 1900' sets his study in the contrasting environments of a rural-agricultural community and an urban-manufacturing one. His findings are set against those of other scholars working with different sources. As the differing environments of the two schools chosen for study would suggest, there is a wide divergence in the causes of absenteeism, largely, it would seem, determined by the very different local customs and the opportunities for children to contribute to the family economy.

Stella Evans's 'A late-flowering logbook: Aylsham community nursery school, 1949-1960' differs from the other articles in this booklet. It covers a much later period than that covered by other participants in the FACHRS school logbook project. This reflects the fact that logbooks are less available, in part because some archivists, the individuals responsible for most of the logbooks that have survived, have refused to make them available under the 100-year rule that applies to census data. Also, the increase in the control of schools by borough and county education authorities, a feature of twentieth-century education policy, led to a standardisation of behaviour that reduced the value of logbooks as a reflection of community economic and social life.[16]

That a primary school is often said to be the heart of a community (usually when it is threatened with closure as with the village shop or post office), may be a cliché, but, as Stella Evans demonstrates, in the case of the Aylsham Community Nursery School, its logbook may well be its voice. From the logbooks at her disposal, Evans, draws out the personalities of the teachers and in some cases their pay and progression; the occupations and roles of parents; the changing impact of advances in the Welfare State, the growth of car ownership, the nature of local shops and tradespeople though their interaction with the school and the changing nature of responsibility for child-care.

Conclusion

It will be apparent from the above and from the transcripts of school logbooks produced by members of FACHRS, that the interest to be derived from their unpredictability is also their weakness. For unlike the standard format of other government initiatives, such as the census enumerators' books or the civil registration of births, marriages and deaths, school logbooks are highly idiosyncratic. They changed when new principal teachers took over and, one suspects, when Her Majesty's Inspectors made their views known. Although some items of information appear more frequently than others, their absence, does not mean that events did not occur, merely that the teacher did not, for whatever reason, feel them to be significant enough to note. It is also apparent from the four articles based on school logbooks appearing below, that, as a source, they are considerably enhanced by the use of other sources.

Notes

1. *Leicester Chronicle* and *Leicestershire Mercury*, 26 May 1900.
2. Robert Lowe, of 'payment by results' fame and, at the time, Vice-President of the Committee of Council on Education.
3. British Parliamentary Papers (BPP) 1862, XLI: 126.
4. BPP 1862, XL1: 232; 277, 321.
5. BPP 1865, XLII: 163-64.
6. BPP 1878-79 XXV: 222.
7. BPP 1867-68 XXV: 434-45.
8. BPP 1867 XII: 190.
9. These are now available on the Members' Only section of the FACHRS website at www.fachrs.com.
10. www.fachrs.com FACH ID 001, School Logbook for Albert Road School, Saltaire, West Yorkshire 1873-74, 6 June 1873, researcher R Coomber.
11. HMI Report on Albert Lane School for 1873 in School Logbook for Albert Road School, Saltaire, West Yorkshire 1873-74, 1 May 1874.
12. *Bury and Norwich Post* and *Suffolk Herald*, 11 March 1862.
13. *Concise Dictionary of National Biography* 1994, Vol I: 500.
14. www.fachrs.com FACH ID 002, School Logbook for Bush Hill Board School (Girls), Enfield, Middlesex, 8 June 1896-27 January 1897, researcher S Hawkes.
15. *Leicester Chronicle* and *Leicestershire Mercury*, 26 May 1900.
16. See the article by Frances Brooks below.

Bibliography

Primary Sources
BPP 1862, XLI. Education minute by the Right Honourable Lords of the Committee of the Privy Council on Education establishing a Revised Code of Regulation.
BPP 1862, XLI. Copies of correspondence addressed to the Lord President of the Council or the Secretary of the Committee of Council on the subject of the Revised Code.
BPP 1865 XLII. Report of the Committee of the Council on Education 1864-65:163-64.

BPP1867-68 XXV: 434-45
BPP 1878-79 XXV: 222

Newspapers
Bury and Norwich Post and *Suffolk Herald,* 11 March 1862.
Leicester Chronicle and *Leicestershire Mercury*, 26 May 1900.

School logbook transcripts produced by members of the School Logbook Project are available to
 FACHRS members via the Members' Only section of the FACHRS website.

Secondary Sources
Oxford University Press. *The Concise Dictionary of National Biography from the Earliest Times
 to 1985*. Oxford University Press, Oxford, 1994.

WINDOW ON COMMUNITY:
THE CHRONOLOGY OF SCHOOL HOLIDAYS 1873-1921
Frances Brooks

This paper will trace why there were changes in the timing of school holidays from the 1870s to the 1920s through the lens of school logsbooks. It focuses on a detailed meticulous case study of a rural Leicestershire Church of England school and this is set in context through a comparison with a number of other schools within the county. It argues that two conclusions can be drawn, firstly that from a period when the timing of holidays varied considerably from place to place, and year to year there developed a gradual standardisation which reached its apogee when control of holidays was vested in Local Education Authorities (LEAs). Secondly, this development marked the changeover from a system that responded to the specific interests of parents, local custom and the economy to one marked by a top-down approach of 'one pattern suits all'. After more than 100 years, there now appears to be a partial return to the earlier system, with the rise of academies and free schools being allowed to determine their own term structures.

Introduction

The use of school logbooks as a source to investigate aspects of social attitude is a neglected field of study but a collection of transcriptions of logbooks from thirty-one schools provides new opportunities for analytical study.[1] In particular it is evident that school holidays were not fixed but had a tendency to vary from year to year. The factors which influenced such fluctuations as yet remain under-researched. It will be argued here that these alterations reflected a number of local factors such as the variations in harvest due to the prevailing weather conditions and social events including feast days, the meeting of the hunt and various local festivities. Thus this article will explore the changing position of holidays within the school year. It will firstly examine the impact of the national church, harvest and feast days on the setting of vacation dates and the length of time allowed focusing on one school. Secondly in order to set this school in the context of Leicestershire and within the wider national picture a comparison will be made with other schools. Finally, there is an extensive historiography surrounding the development of a school based education and this article will not seek to add to this debate but will instead investigate how school holidays developed and the rationale behind the system. School logbooks themselves were intended to provide a regular record of the activity of a school and were a requirement of the 1871 New Code which stated that:

the briefest entry which will suffice to specify either ordinary progress, or whatever other fact concerning the school or its teachers, such as the dates of withdrawals, commencements of duty, cautions, illness, etc., may require to be referred to at a future time, or may otherwise deserve to be recorded.

This study will concentrate on Cosby a small village in Leicestershire, about six miles south east of Leicester; the school was built through public subscription in 1872 at the instigation of the vicar. It served a community that had almost doubled in the first half of the nineteenth century from 555 in 1801 to 1,026 in 1851 but declined to 944 in 1871. At the latter date the adult male population was split between agriculture (29%) and framework knitting (21%), with most of the rest in service industries and a nascent boot and shoe industry (4%). Both agriculture and framework knitting were family-based domestic industries in which children played a significant part. Given that framework knitting was in decline (it disappeared completely over the next 50 years), those attached to it were likely to be eager for the extra earnings available at harvest time. By 1911 the population had increased to 1,560 inhabitants. Pamela Horn has revealed the importance of the school logbooks as a source of information on the wider community with frequent mentions of irregular attendance due to the need for children to work, communal events and national celebrations.[2] Whilst extracts from logbooks have provided interesting insights into school life[3] it is case studies such as William Marsden's study of Fleet Road Board School[4] and Susannah Wright's comparison of urban schools in Leicester and Birmingham[5] that provide an opportunity to systematically study the activities in schools and the communities they served. In this context 'community' is taken to mean the families within the catchment of the school. In a village such as Cosby this was usually the whole village, even though that may contain smaller sub-communities based on other criteria such as place of worship. In more remote areas this might incorporate families from several villages and in an urban area there may be several school communities within a given parish.[6] A unifying aspect of social cohesion has always been the customary calendar of feasts and holidays and prior to the mid nineteenth century there was widespread support for customary law, whether general or local and many of the local traditions had their roots in dole days and other common rights held between landowners and the labouring poor.[7] Others, relating to the monarchy, were legitimised by their earlier inclusion in the Church of England Book of Common Prayer. These included the Martyrdom of King Charles 1 on 30 January; 29 May, which as Oak Apple Day, celebrated the restoration of the monarchy; and Guy Fawkes Day on 5 November.[8] Such State services had been abolished in 1859, which contributed to undermining their legitimacy.[9] As the century progressed many of the local fairs and festivities came under increasing criticism and Horn believes the 1871 Act that facilitated the closure of fairs and was 'symptomatic of the hostility and prudishness of many of the better-off towards these entertainments'.[10] There were widespread attempts to curtail the excesses of dole days, with even Christmas boxes being banned in some place.[11] It has been argued that Whitsun was a traditional, frequently drunken, festivity which in the nineteenth century was brought under control through its adoption by friendly societies as an excursion or activity day.[12] In many places, including Cosby, the Band of Hope were quite successful in taking over the new August Bank holiday, which had been introduced in 1873 as the first official purely secular holiday, and organising outings to meetings and rallies.

Logbooks can even indicate variations in regional weather patterns as the timing of harvest is a frequent determining factor in holiday date, certainly in the nineteenth century. The detailed, if generalised, notes given in JM Stratton's *Agricultural Records* are reflected in the timing of holidays in Cosby. Noticeably, in 1887, which Stratton recorded as an unusually dry year with good harvests, the holiday was taken from 12 August to 12 September but in the following year, which was recorded as rather wet with poor harvests the Cosby harvest holiday was not taken until 7 September and lasted until 8 October

> As only 44 children came to school this morning the children were dismissed for harvest holidays.[13]   (Cosby SLB: 10 September 1883)

> School reopens after the Holidays but numbers so low it was thought best to continue the Holidays and as the village feast is next week the school will not commence till Monday October 22nd.[14]

The agency revealed by the children (or, more probably, their parents), in the above quotations from a late nineteenth century school logbook cannot but appear surprising to the generations of school children, and their parents, who have endured top-down decision making as to the length and timing of school holidays for over a century. The detailed analysis of one Church of England school in Leicestershire shows that factors affecting the time and duration of non-school days were rather different in the nineteenth century.

National Church, Harvest and Feast Holidays

When Cosby School opened in October 1872 the first holiday given was recorded thus 'On Friday Morning the school was closed for week it being Christmas Time'; when they returned, the next week's entry included the comment that 'On Wednesday (being New Year's Day) the school was closed'.[15] Until the turn of the century the Christmas holiday was normally taken from Monday to Friday of the week including Christmas Day but by 1903 this was extended to include the first week of the New Year. In the school's opening year only Good Friday and the Easter Monday were taken at Eastertide, but in 1874 the holiday was extended to Good Friday and the whole of the next week and this became the accepted situation until after the Local Education Authority (LEA) took over in 1903, when the break was generally extended from six to eight days. The pupils also had one week's holiday for Whitsuntide, occurring seven weeks after Easter, and this was also increased by an extra day after 1906.

From the evidence of the school logbook it is clear that in the latter half of the nineteenth century the timing and length of the holiday taken in the summer and early autumn months varied considerably and was adjusted to meet the needs of the farming community. Figure 1 shows that in the summer of 1873 (week 30 is around the end of July) the school holidays extend from 18 August to 12 September. In the following year a similar four-week holiday started two weeks earlier on 3 August and then in 1875 the period moved to a later date, starting on 20 August. This variation does indicate the date of the harvest as it reflects the observations of Stratton who records the harvests of those years as poor, good and poor, respectively.[16] Three years in which Stratton observed very wet weather the Cosby holiday extended right into October; in 1879 it ended on 10 October, in 1883 on 12 October and in 1888 on 5 October.

Although agriculture was still a major part of village life after 1895 the variation seen in the earlier years stabilised and settled on a period of four weeks starting between 27 July and 7 August. An entry in the logbook of 30 July 1897 however suggests that the timing of the harvest still had, at least some, influence. The entry reads: 'Attendance good but many children away taking meals to their parents in the Hay and Harvest field'.[17] Whether this move away from the natural cycle of nature was due to a decline in the population's dependence on agriculture for employment (wholly or in part) through increased mechanisation is open to question but the number of labourers required to bring in the harvest fell by 80% by the time steam machinery was introduced around 1900.[18] Other social factors relating to the supposed need to impose order and control over the lives of the labouring classes also had some bearing. This study shows that standardisation from year to year in both the length and timing of the summer holiday predates the setting up of Local Education Authorities in 1903 after which they set the date of holidays in all schools in the locality. During WWI the shortage of labour on the land was a national problem and schools were encouraged to help with the war effort. This was demonstrated from an entry in the logbook for1815 which recorded:

> School re-opened after five weeks holiday. The 5th week was given to enable some of the elder children to assist in harvesting where necessary.[19]

In 1917 some of the boys were allowed time out of school 'to dig in the gardens of men away with the forces'.[20]

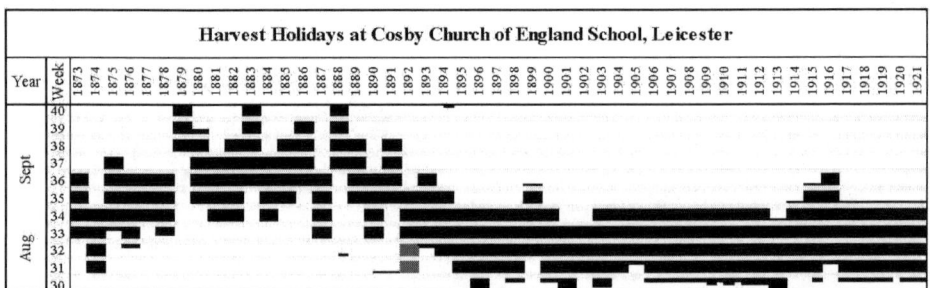

Fig. 1. Timing of harvest holidays at Cosby Church of England School 1873-1921

An important feature in Cosby village life was the annual feast and the timing of this reveals the persistence of old customs in some villages. The church was dedicated to St. Michael and All Angels and the feast associated with it was taken the week after the 'old' Michaelmas Day, 11 October.[21] Unlike holidays celebrated nationally the length of the school closure associated with the feast was reduced from a week in the nineteenth century to just two days when the LEA took control although the traditional timing was retained.

These holidays are familiar to anyone educated in England in the twentieth century but in Cosby, as in many places, additional half days were regularly given on numerous occasions for events such as Shrove Tuesday and church teas, etc. Shrove Tuesday was traditionally celebrated in many Leicestershire villages when the church bells were rung at noon, originally to call parishioners to confession before Lent. In time it became less a call to church and more a reminder to make pancakes and marked the start of the local festivities. In Cosby the Shroving bell is recorded as still being rung in 1876.[22] The village seemed reluctant to give up this tradition. 'Attendance low Shrove Tuesday, usually a Half Holiday but Holiday last Thursday instead'.[23] The tradition of "Whipping Toms", a rather barbaric combat, which had taken place on Shrove Tuesday around The Newark, the precincts of Leicester Castle, since the Middle Ages had been banned by the middle of the nineteenth century; however the tradition for revelry was still strong in the town and surrounding villages where the half-holiday continued to be given throughout this period.[24]

That concept that elementary education should have a religious dimension had been accepted by the time of the Forster Act of 1870 and as a Church of England school the Cosby vicar took an active role and was in attendance several times a week.[25] The school also served the children of families belonging to the Methodist and Baptist churches in the village but until 1882 a half-day holiday was given only for the Church of England treat. From the mid 1880s the Church treat was increased to a full day's holiday, but both the Methodists and the Baptists started holding teas or treats on school days, which resulted in low attendance. It was noted, in 1891, 1898 and 1901, in the school logbook that the afternoon registers were not marked on these days; it seems there was an advantage when reporting attendance figures in 'giving a holiday' rather than having very low attendance. From 1906 under the influence of the LEA there was evidence of more interdenominational equality as all three churches held teas or treats during the summer term for which half holidays were given for all the children regardless of their religious affiliation. There is less evidence of such treats after the start of the Great War when changing attitudes toward school attendance encouraged such events to be then held during the summer break. For a short period Empire Day, introduced in 1916, was the occasion for a half day holiday but later it was celebrated within the school rather than in the village although it remained a popular cause for celebration for over fifty years.[26]

In Cosby in the late 1880s August Bank Holiday was sometimes the occasion for Band of Hope or Temperance League rallies, for which the school would close, if not the attendance was usually low or, in 1888 for example, the afternoon would be given as a holiday and it was recorded:

> Attendance not taken Monday afternoon as several were taken by their friends to Leicester and other places being Bank Holiday and many Excursions. The few who came were dismissed. (Cosby SLB: 10 August 1888).

The following year the school again acknowledged the reality of the situation and the afternoon was considered a holiday, even though some children had attended:

> Monday numbers low owing to Flower Show at Croft & Bank Holiday. In the afternoon attendance not taken children dismissed at 3pm. Excursion to Grimsby of Mr Holyoak's men on Tuesday with the children. The above reasons account for lower average this week.[27]

Although much has been made of May Day celebrations with the suggestion that it was widely given as a holiday when processions would take place and children would dance around the Maypole. Where this did not occur children would often take the day to wear May garlands however there are no references to any celebrations in Cosby logbooks.[28]

Comparison of summer holidays in other places

Although the experience of the local impact of economic circumstances, notably the timing of the harvest, in Cosby is borne out by the evidence presented above, the question still remains as to the use of school logbooks as a surrogate measure of local customs, notably those impacting on school holidays. In order to widen the range of study, a 5% sample was drawn of the 260 schools listed in Kelly's 1881 *Directory for Leicestershire*. The aim was to reflect the governing bodies (58% were Church of England or National Schools, 19% were locally endowed and 13% were Board Schools) as well as the urban/rural split (Figs 2-5). Three subjects were taken for purposes of comparison: the annual number of days holiday (Fig. 2); the variation of school holidays between rural and urban areas (Fig. 3); the variation of school holidays in Church and Board schools (Fig. 4) and the timing of Leicestershire school holidays in 1881 and 1921 (Fig. 5).

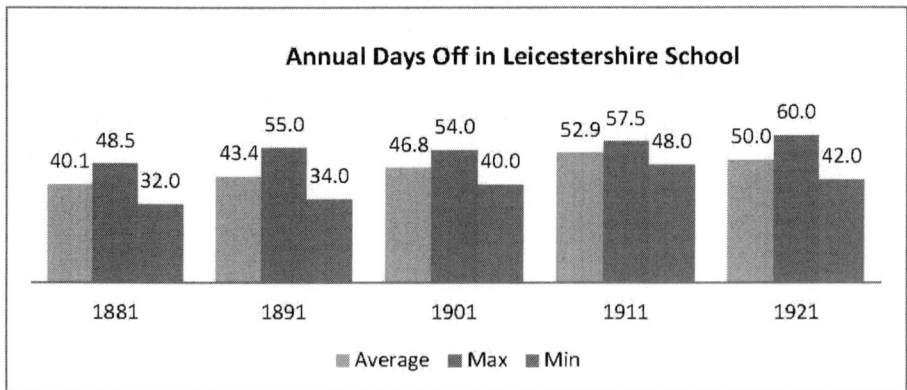

Fig. 2. The average, maximum and minimum number of days' holiday in a sample of Leicestershire elementary schools 1881-1921

Figure 2 shows that the mean number of day's holiday in the sample schools grew fairly regularly from forty to fifty days between 1881 and 1921. The jump in 1911 is an anomaly as an extra week's holiday was given to celebrate the coronation of George V. The difference between the schools with the shortest holidays and those with the longest remained considerable and does not appear to change in any regular way. Grace Dieu, a Roman Catholic school, had as many as sixty days holiday in 1921. This included five Saints days that were not usually given at other schools. In the same year the Church of England school at Burton Ovary had only forty-two days' holiday.

The holiday at Easter also shows a tendency for gradual standardisation and lengthening of the break over the period 1881-1921. In 1881 the break ranged from half a day to eight days with six days being the most popular duration. In 1881, in the Church of England school in Lusby (Lincolnshire) only a half day was given on Good Friday and in 1891 there was no mention of any break at all at Easter. This may be a limitation of the school logbook as a source as, in this case, the entries are confined to the average weekly attendance. There was a gradual increase in the length of the holiday at Lusby to two days in the early twentieth century and to six days by 1921. Only one Wesleyan school appeared in the Leicestershire sample, that at Coalville. Their holidays were very similar in both timing and length to those of the other religious schools in the county. However three Wesleyan schools examined in Lincolnshire all took very limited holidays at Easter until the twentieth century and that at Market Rasen rarely recorded any holiday at that time.

In Leicestershire, the number of days given as holiday over Christmas and the New Year period also increased slightly for most schools, the main change being the addition of a week's holiday at New Year. In Cosby this was taken by 1905 and in all other schools in the Leicestershire sample by 1911. The one exception was at Burton Ovary in 1921 when the school 'opened on 3 January to make up necessary attendances'.[29]

It is apparent from Fig. 3 that the average number of days' summer holiday in Leicestershire's urban areas was considerably less than in its rural areas until the early years of the twentieth century and that the difference persisted until at least 1921. In Leicester there was no provision at all for summer holidays in the Board Schools in 1875. The head teachers had petitioned to close their schools during their own two-week summer holidays but the School Board considered this too great an inconvenience for parents who worked while their children were in school and thought it 'would be an injury to the children both morally and educationally, and inflict a loss on the ratepayers'.[30]

There was also a difference in the length of school holidays between the Board Schools and the Church Schools in the Leicestershire sample (Fig. 5). In 1881, 1891 and 1901, the Church Schools took longer holidays than the Board Schools although the difference was, on average, small. The difference does not appear to have been due to extra religious holidays as many Church schools preferred to have the school open on Holy Days, such as Ash Wednesday when the pupils could be escorted to church and thus ensure attendance. In 1911 and 1921, the difference was reversed with both types of school taking longer holidays than previously.

In 1881 the timing of the main summer holiday showed marked variations with St Marks National school, in the Belgrave suburb of Leicester, returning from their two-and-a-half-week holiday on 18 July even though the Belgrave Feast was held that week resulting in poor school attendance. At the other extreme Burton Ovary National school did not start their five-week holiday until 22 August. Copt Oak National School, with just fifty pupils, split its holiday between the summer and autumn harvest times. The startling difference between rural and urban school seen in 1881 has been standardised by 1921 (Fig. 5) and the days of some schools starting their summer holidays at the beginning of July, whilst others delayed until September was over, with ten of the twelve schools starting their summer holidays in the first week of August. The exceptions were the two Leicester schools, which started their holidays in the last week of July.

St Michael's was clearly a local feast as other villages in the area held their Feast or Wakes weeks on different dates and Cosby children were often absent to enjoy the neighbouring festivities: 'Whetstone Wake attendance on that account low on Monday'.[31] Broughton Astley Wake was also popular with Cosby children and it often fell within their term time: 'Attendance very low on Monday; some of the children gone to Broughton Wake'.[32] However, Broughton Astley School incorporated the feast into their delayed summer break, in 1871 and 1881 they were off until the middle of September and even in 1901 they did not return to school until 9th September.[33] Of the Leicestershire schools considered only Castle Donnington took a regular holiday for their feast, which was taken two weeks later than Cosby's. In 1910 the Leicester LEA resolved that all schools would have an autumn holiday from midday 28 October to 2 November.[34] In 1919 when King George V suggested that some extension of the school holidays be given in commemoration of the peace the council decided that no extension should be given to the summer holiday but the autumn break could be increased from two to four days and taken at the beginning of October instead of at the 'end of the month as provided by the regulations'.[35] In the event this was changed to a two days' extension to the Midsummer

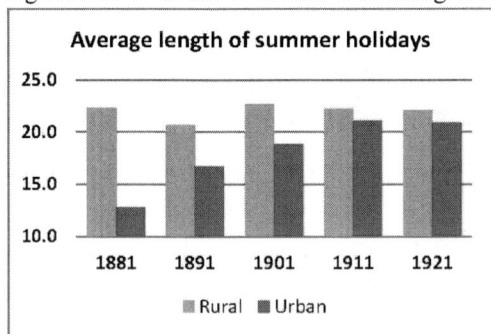

Average length of summer holidays

Holidays to fall in line with the Secondary Schools. By 1921 all the Leicestershire schools surveyed took an autumn break at the end of October. However, this had not become universal across the country by this date.

Fig. 3 The average length of summer holidays in rural and urban schools in Leicestershire 1881-1921

The timing of feasts, fairs and wakes weeks varied widely across the country. In Lincolnshire most were held in the spring with Lincoln Fair being held towards the end of April when St Martin's Church of England School, Lincoln, had a week's holiday for the fair in addition to Easter. The fair then moved on to Boston where the charter fair, dating back to 1573, was proclaimed on May 3 every year. In the second half of the nineteenth century, although it was still the occasion for servant hiring's, the traditional trading fair was losing significance and had been largely replaced by the pleasure fair, which was a popular event in the wider area with omnibuses and trains bringing visitors from all over Lincolnshire. St Botolph's National School in Boston had at least the five days off for May Week throughout the 1881-1921 period. Even Kirton Holme Board

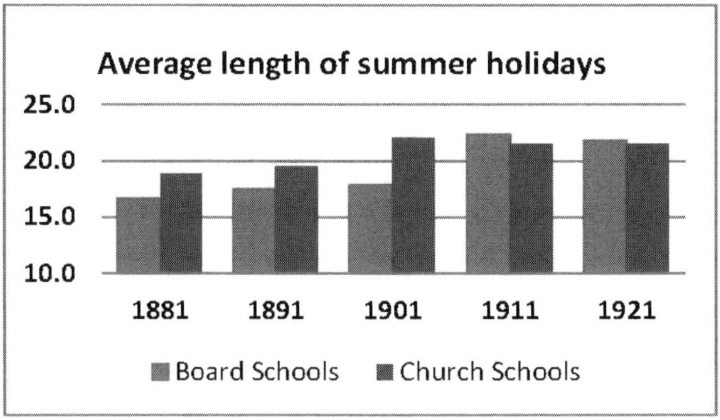

Average length of summer holidays

| | 1881 | 1891 | 1901 | 1911 | 1921 |

■ Board Schools ■ Church Schools

Fig. 4. The average length of summer holidays in Board Schools and Church Schools in Leicestershire 1881-1921

School, some five miles outside Boston, were free to visit the fair for two days a year until 1901 when the practice was cancelled. However, even by 1921 no record has been found of an autumn mid-term break in a Lincolnshire school.

In his study of feasts, in the mining villages of South Yorkshire, Walker found that the September feast week was frequently celebrated as part of the school holiday but by 1888 moves were made to change the nature of the feast, reducing the bawdy element in favour of sporting events and charitable fund raising. At this period attempts were made to move the feast to August, a slack time for the collieries, with inducements made to the colliers if they would work just two days of the feast week but it seems local tradition won the day with the feast reverting to its customary September date.[36]

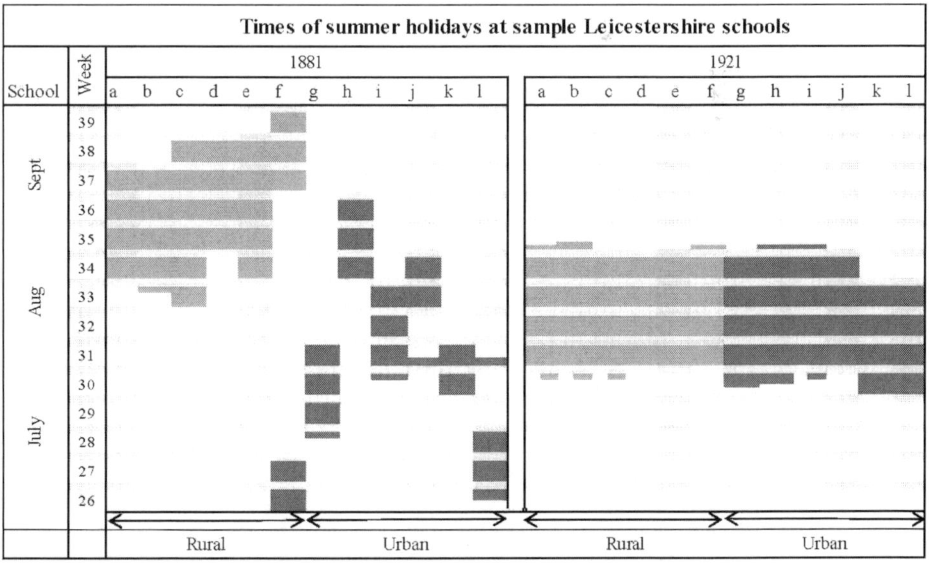

Fig. 5. The timing of summer holidays in a sample of Leicestershire schools in 1881 and 1921

The Huddersfield School Board (West Riding of Yorkshire) decided in 1873 on a three weeks' summer holiday for one group of schools under its control plus another week during Honley Feast (late September).[37] Honley Feast was the last of the season of feasts held in the area and often drew large crowds, many coming by train. At its core were 'beef, beer and red cabbage', but it also featured cricket matches, races, a fair, circus and sideshows. It received a report annually in the local paper and its concomitants – various drunken brawls leading to police prosecutions – were also reported. By 1888 the official week-long holiday had been reduced to two days – this may reflect a decline in the popularity of the feast with a consequent diminution in its length or may have been an imposed reduction to control excessive rowdy behaviour. By 1889 in some parts of the West Riding no holidays were allowed for the local feast and when an appeal was made to the Huddersfield School Board relating to the Longwood Feast it was refused. The Headmistress of the Oakes School had asked that absences during the Feast should 'not disqualify children who would otherwise have earned attendance prizes'.[38] It is a limitation of this study that due to data protection restrictions many logbooks relating to the twentieth century are not available so alternative sources are required to ascertain when an autumn break became widespread.

Attitudes to education

The introduction of compulsory schooling also gave children the right to an education but this has not always been fully appreciated. Parents frequently resented the cost of education and the loss of potential income by their children when at school. When the availability of free schooling was announced in 1891 Cosby School managers were not inclined to take up the offer, fearing loss of control, but were forced to accept the financial support offered when parents withdrew their children:

> Gleaning keeps some away owing to the late harvest. But the chief cause of the greatly reduced attendance is that the children are not allowed to attend without their school fees. As the manager at a meeting held on August 31st declined the fee grant for the present'.[39]

It is noticeable that after this date less consideration was given to the communities and concessions to the needs of the harvest. Employment of children less than ten years of age had been banned under the 1878 Factory Act, but Cosby had a tradition of children contributing to the household economy and although framework knitting had declined in the village by 1891 advantage was taken of any opportunity for casual work. The opening of the nearby Narborough Golf Club meant that from 1896 attendance became a frequent problem on Thursdays. 'Only two boys in the first-class and many away from the other classes at the golf field'.[40] This was not a simple case of boys skiving off in search of lost balls but a major problem, condoned by golfers of the village requiring caddies. 'Mr W. J. Clarke Manager of the school asked for 10 boys to carry clubs at the golf match on Thursday afternoon, permission was given to 10 who were at school but there were many others who did not attend'.[41] This was a problem, which, despite the efforts of the Master and the Attendance Officer, persisted until 1905. It has been recorded that as late as 1901 in some areas of Yorkshire the requirements of the Education Act were suspended for three weeks, or longer, to allow children to work in the fields. If holidays were not granted children would just stay away when paid employment was available. This attitude was encouraged by employers who saw little need for the children of labouring families to be educated.[42] The fact that school closures were frequently allowed for church and chapel treats is an example of how religious organisations continued to demonstrate their control over the community – in other words children must attend unless we deem otherwise.

The problem of schools choosing to give holidays at different times had been highlighted as early as 1864, so far as Public Schools were concerned. The Royal Commission on Public Schools of 1864 found that most schools took either two or three holidays a year varying in total from fourteen to sixteen weeks. This was considered inconvenient for families with boys at different schools and some standardisation was suggested.[43] In the mid nineteenth century the pattern of holidays accommodated the needs and traditions of the families they served but also included breaks for the Church festivals of Easter and Christmas with the mid-term breaks at Whitsuntide and in some areas the feast week in the autumn. It is evident from the Cosby logbooks that other local schools did not take the same holiday dates, and Cosby's children were often absent, attending other local feasts. In 1912 a secondary school was opened in Cosby with handicraft and cookery facilities, which were used by pupils from neighbouring villages. The problem of varying holidays is evidenced from an entry in the logbook: 'The Cookery Class did not meet owing to Narborough feast. The Narborough school contingent would have been absent'.[44] Under the auspices of the LEAs in the years following the First World War standardisation was largely achieved in all schools under their control and has remained fairly static until recently when the introduction of Academies and Free Schools is likely to lead to a reduced role for the Local Education Authorities.

Conclusion

Use of school logbooks for comparative research has serious limitations as the data recorded varied according to the whim of the recorder. These variations however can tell as much as they conceal, highlighting the perceived relative importance of academic achievement, pupil behaviour and attendance or extracurricular activities, such as Sunday School treats and local customs. Analysis of the case school has demonstrated that although communities were able to exert a strong influence on the way that legislation was put into practice at the local level in the early years of compulsory education, this right was gradually forfeited even before control was passed up to Local Education Authorities. Some of this general standardisation was due to the influence of the state but attempts at local social control curtailed many of the local bacchanalian festivities. Comparison of holidays between schools also shows the tendency over time to reduce the variation between schools in urban and rural communities and between Boards schools and those with religious affiliations. There is scope for more extensive research to consider schools in other situations, such as where part timers were prevalent or where maternal employment outside the home was the norm. School logbooks can provide a unique perspective on the lives of the communities they serve although there are limitations. The range of data recorded is not consistent, even holidays are not necessarily recorded, possibly because that data would be available through attendance registers. Not all logbooks have survived and stricter application of data protection has reduced access to many. It is apparent that the pattern of holiday, which had been established by the second decade of the nineteenth century, remained largely unchanged for nearly a hundred years. Whether or not the agency given to Academies and Free Schools in recent times to determine their own structures of school timetabling will lead to significant changes and whether these reflect modern facets of community only time will tell. We may see change again on the horizon as increasing numbers of families do not have a non-working parent or extended family to provide child care and it may fall to schools to provide affordable support for longer hours and more days of the year. In recent years there have been state attempts to prevent absenteeism due to term time holidays through fines and prosecution but a judicial review in May 2016 may have passed some of these rights back to the family.[45] The economic problem of peak prices during school vacations and the resultant absenteeism may be tackled by spreading the timings of holidays. However, this would reintroduce the problems that standardisation was intended to mitigate. Either of these scenarios could be seen as a return to some community control over schools to reflect the employment and social needs of the community.

Notes

1. These arise from a mini-project undertaken by the Family and Community Historical Research Society on school logbooks.
2. Horn, P. 'School Log Books' in *Short Guides to Records*, Thompson, KM (ed.)Second Series, London: Historical Association, 1997, 105.
3. Gant, R. 'School records, family migration and community history: Insights from Sudbrook and the construction of the Severn Tunnel.' *Family & Community History*, 2008, 11:1, pp.35-.6.
4 *See* Marsden, WE. *Educating the Respectable: A Study of Fleet Road Board School Hampstead, 1879-1903*. Woburn Press, London, 1991.
5. *Refer to* Wright, S. "Teachers, family and community in the urban elementary school: evidence from English school logs books c. 1880-1918.' *History of Education*, 2012, 41:2, pp.155-73.
6. *See* Mills, D. 'Defining community: a critical review of 'community' in Family and Community History'. *Family & Community History*, 2008, 7:1, pp.7-8.
7. Bushaway, B. *By Rite: Custom, ceremony and community in England 1700-1880*. Breviary Stuff Publications, London, 2011, p.6.
8. Ibid., p.41.
9. Ibid., p.163.
10. Horn, P. *Pleasures & Pastimes in Victorian Britain*, Sutton, Stroud, 1999, p.84.
11. Bushaway, p.173.
12. Reay, B. *Rural England: labouring lives in the nineteenth century*. Palgrave Macmillan, Basingstoke, 2004, p.121.
13. Leicestershire County Record Office (LCRO), E/LB/78/1-3, Cosby School logbooks, 10 September 1883.
14. Ibid., 8 October 1883.
15. Ibid., 20 December 1872-3 January 1873.
16. Stratton, JM. *Agricultural Records A.D. 220-1968*. Baker, London, 1978, p.83.
17. LCRO, E/LB/78/1-3, Cosby School logbooks, 30 July 1897.
18. Bushaway, p.68.

19. LCRO, E/LB/78/1-3, Cosby School logbooks, 7 September 1915.
20. LCRO, E/LB/78/1-3, Cosby School logbooks, 2 May 1917.
21. Michaelmas was moved to 29 September with the revised calendar of 1752, however Cosby was one of the places that kept to the old dates.
22. Bilson, CJ. *County Folklore of Leicestershire & Rutland*. Nutt for Folklore Society, London, 1895, p.72.
23. LCRO, E/LB/78/1-3, Cosby School logbooks, 5 March 1897.
24. Bushaway, p.161.
25. Best, GFA. 'The Religious difficulties of national education in England, 1800-1870'. *The Cambridge Historical Journal*, 1956, 12:2, p.156.
26. English, J. 'Empire Day in Britain 1904-1958'. *The Historical Journal*, 2006, 49:1, p.249.
27. LCRO, E/LB/78/1-3, Cosby School logbooks, 8 August 1889.
28. Horn, P. *The Victorian Country Child*. Sutton, Stroud, 1997, pp.177-82.
29. LCRO, E/LB/51/1-2, Burton Overy School logbook, January 1921.
30. *Leicester Chronicle*, 29 May 1875, p.7.
31. LCRO, E/LB/78/1-3, Cosby School logbooks, 10 October 1884.
32. LCRO, E/LB/78/1-3, Cosby School logbooks, 6 September 1883.
33. LCRO, E/LB/51/1-2. Broughton Astley, School logbook, 9 September 1901.
34. LCRO Leicester Elementary Schools' sub-committee Minutes 37, p.70
35. LCRO, Leicester Elementary Schools' sub-committee Minutes 40, p.56..
36. Walker, A. 'Feasting in a South Yorkshire colliery district: resistance and accommodation to customary change in Wombwell and Darfield, c1860-1900' *Family & Community History*, 2001, 4:1, p.9.
37. *The Huddersfield Daily Chronicle*, 4 November 1873.
38. Ibid., 5 March 1889.
39. LCRO, E/LB/78/1-3, Cosby School logbooks, 2 October 1891.
40. Ibid., 12 May 1899.
41. Ibid, 2 November 1900.
42. Horn, P. *The Victorian & Edwardian Schoolchild*. Amberley, Gloucester, 2010, pp.106-7.
43. Pimlott, JAR. *The Englishman's holiday: a social history*. Harvester Press, Hassock, 1976, p.158.
44. LCRO, E/LB/78/1-3, Cosby School logbooks, 6 November 1916.
45. *The Guardian*, 13 May 2016.

Bibliography

Primary Sources
Leicestershire County Record Office
School logbooks (SLBs)

a. Cosby	E/LB/78/1-3
b. Broughton Astley	E/LB/51/1-2
c. Somersby	E/LB/297/1-3
d. Sutton Cheney	E/LB/314/1-2
e. Burton Overy	E/LB/59/1-2
f. Copt Oak	E/LB/216/1
g. Castle Donnington	DE5569/1,3
h. Grace Dieu, Whitick	DE4503/4-5
i. Coalville	E/LB/73/1-2

j. Oadby	E/LB/241/1-3
k.Catherine Street, Leicester	DE2938/1-2
l. St Marks, Leicester	DE3893/23-24

Kelly's Directory: Leicestershire 1881.
Leicester Elementary Schools' sub-committee Minutes 19D59/VII/40:56

Lincolnshire County Record Office
School logbooks (SLBs)

St Botolph's, Boston	Par/16/8
Kirton Holme	SR521/8/1-2
St Martin's, Lincoln	SR626/8/1-2
Lusby	SR 712/8/1-2
Market Rasen	SR 731/8/1-2

Secondary Sources

Best, GFA. 'The Religious difficulties of national education in England, 1800-1870' *The Cambridge Historical Journal*, 1956, 12:2, pp.155-173.

Bilson, CJ. *County Folklore of Leicestershire & Rutland*. London: Nutt for Folklore Society, 1895.

Bushaway, Bob. *By Rite: Custom, ceremony and community in England 1700-1880*. Breviary Stuff Publications, London, 2011.

English, J. 'Empire Day in Britain 1904 – 1958' *The Historical Journal*, 2006, 49:1, pp.247-76.

Gant, R. (2008) 'School records, family migration and community history: Insights from Sudbrook and the construction of the Severn Tunnel.' *Family & Community History*, 2008, 11:1, pp.27-44.

Horn, P. 'School Log Books' in *Short Guides to Records*, Thompson KM (ed.), Second Series, Historical Association, London 1997a.

Horn, P. *The Victorian Country Child*. Sutton, Stroud, 1997b.

Horn, P. *Pleasures & Pastimes in Victorian Britain*. Sutton, Stroud, 1999.

Horn, P. *The Victorian & Edwardian Schoolchild*. Amberley, Gloucester, 2010.

Langley, A. *Victorian Village Life*. Stretton Millennium History Group, Stretton on Dunsmore, Warks, 2004.

Lewis Shipman, Lillian. (1973) 'The Band of Hope movement: Respectable recreation for working-class children'. *Victorian Studies,* 1973, 17:1, pp.49-74.

Mills, D. 'Defining community: a critical review of 'community' in Family and Community History'. *Family & Community History*, 2004, 7:1, pp.7-8.

Marsden, WE. *Educating the Respectable: A Study of Fleet Road Board School, Hampstead, 1879-1903*. Woburn Press, London, 1991.

Pimlott, JAR. *The Englishman's holiday: a social history*. Harvester Press Hassock, 1976.

Reay, Barry. *Rural England: labouring lives in the nineteenth century*. Palgrave Macmillan, Basingstoke, 2004.

Stratton, JM. *Agricultural Records A.D. 220-1968*. Baker, London, 1978.

Walker, Andrew. (2001) 'Feasting in a South Yorkshire colliery district: resistance and accommodation to customary change in Wombwell and Darfield, c1860-1900' *Family & Community History*, 2001, 4:1, pp.5-18.

Wright, S. (2012) 'Teachers, family and community in the urban elementary school: evidence from English school logs books c. 1880-1918.' *History of Education*, 2012, 41:2, pp.155-73.

Digital Sources
British Library: 19th century newspapers
Huddersfield Daily Chronicle
Leicester Chronicle
www.theguardian.com/politics/2016/may/13
www.fachrs.com: OU DA301 Report 1995, Brooks, F. 'Household Structures and Family Economy in Cosby, Leicestershire. 1851-1891', 1994.
www.fachrs.com: SLB Project reports

Acknowledgements

My thanks to members of FACHRS who have shared their research findings from the 'School and Community Project'

Biographical note

Frances Brooks is a graduate of the Open University including Studying Family and Community History: 19th and 20th centuries for which she completed a project: Household Structures and Family Economy in Cosby, Leicestershire. 1851-1891. She is currently treasurer of the Family and Community Historical Research Society.

JOBS FOR THE GIRLS: THE RECRUITMENT OF TEACHERS IN WINCHCOMBE, GLOUCESTERSHIRE (1860-1910)
Robert W White

This article explores how teachers were recruited at the national/board schools in the parish of Winchcombe and the extent to which their engagements may have been influenced by nepotism. The documentary evidence for the recruitment of school staff in late nineteenth century Winchcombe is sparse, but anecdotal evidence from school logbooks, personal diaries and other sources, shows changes in recruitment methods during this fifty year span. In addition, detailed family reconstitution has been carried out, using parish registers and the decennial census enumerators' books, in order to determine the family links between teachers in Winchcombe's schools.

Introduction

The social historian, Pamela Horn, referred to 'the favouritism which some [school] masters and mistresses showed towards the children of the larger farmers or the squire's agent as opposed to those of the labourers or the village craftsmen'.[1] This partiality was said to be regarding the use of the cane (which was rarely used against the favoured few) and, during the winter months, the proximity of the favoured children to the single fire in the schoolroom. But another type of favouritism may have been evident in the selection of pupils to become teachers.

The new Whig government of 1846 decided to enhance educational standards by improving teachers qualifications and replacing the old monitorial style of teaching with a more proficient pupil-teacher system, and Sir James Kay-Shuttleworth, Permanent Secretary of the Committee of Council on Education, argued that pupil teachers would 'for the most part belong to families supported by manual labour'.[2] One historian has indicated that in the nineteenth century 'elementary education opened a door for intelligent poor girls and a handful squeezed through it'.[3] It provided an opportunity for them to 'better themselves' as well as gain some independence, because qualifying as a teacher often meant applying for a position in another location and thus involved leaving home to live and work elsewhere. In 1859 there were 6,605 male and 6,253 female pupil teachers in England and Wales, but by 1896 the numbers had increased to 7,737 males and 28,137 females, the reason for this disparity being that there were more, and often better paid, employment opportunities for boys.[4]

So, during the second half of the nineteenth century elementary school teaching gradually became a female-dominated occupation because women were far cheaper to employ, but being a pupil teacher was by no means an easy option because they spent much time in private study as well as teaching a class of up to fifty children. In addition, they received tuition in out-of-school hours from the schoolmistress at a time convenient to her – hence in the 1860s the pupil teachers at Winchcombe Girls School commenced lessons with the mistress at 6.45am.[5] Also, unlike most other occupations undertaken by young people, teachers were in the public eye. By the 1870s some local newspapers were sending reporters to school board meetings, so not only were the examination results of pupil teachers sometimes published in local newspapers but so were the names of those who did not make the grade. Hence, in 1879, Alfred Smith, aged 15, was said to have 'tried the duties of assistant monitor at Winchcombe Boys School but felt he must decline them'.[6]

Winchcombe and its first national school

Winchcombe is a large rural parish in north Gloucestershire dominated by the small market town of Winchcombe. The rest of the parish consists of a series of rural hamlets, the largest of which is Gretton. By the 1830s agriculture was still the dominant industry in the parish with almost half the male workforce being farmers or agricultural labourers. Other than agriculture, what little industry Winchcombe had was centred on the River Isbourne, which runs through the town, notably a paper mill and also flour and grist mills, plus a silk mill and a tannery. The population of Winchcombe parish in 1861 was 2,937 and remained fairly constant over the next fifty years.[7]

Illustration I. *Winchcombe's first National School, opened in 1857 Photo © Alastair Robinson*

The first National School in Winchcombe, taking girls, boys and infants, was opened in January 1857 with the construction of a new school building – later known as the Assembly Rooms – in Abbey Terrace, funded by William Smith, a local solicitor.[8] A separate establishment known as Dents School was financed by John Coucher Dent of nearby Sudeley Castle and opened in January 1868.[9] This was initially for girls and boys – with the infants remaining in the Assembly Rooms – but when a Boys School was set up in 1876 Dents School became the Girls School.[10] By then another school had been established in the hamlet of Gretton, in 1861, two miles distant, taking girls and boys of all ages.[11] Thus, by the mid 1870s Winchcombe parish was perhaps unusual in having four elementary schools, three of them being under the control of the newly-created Winchcombe United School Board, although Dents School was managed by a separate board of trustees.[12]

School logbooks – introduced nationally in 1862 for completion by each head teacher[13] – reveal that pupil teachers and monitors were employed in all four elementary schools in Winchcombe parish, and over the next fifty years a total of 139 pupil teachers and monitors, most of them local girls, taught there. They were often recruited straight from school, or a year or two after leaving, and were usually aged about thirteen or fourteen when they commenced work. Pupil teachers, often being former pupils of the same school, were indentured for five years – reduced to four years in 1877 – at the end of which time the most able were offered Queens Scholarships to take a two-year course of study at a training college.[14] Less able pupil teachers could continue to teach whilst studying at home, taking external examinations with the same end in sight. Like the scholars in their charge, the pupil teachers were examined each year by a government Inspector (HMI) and, until 1862, if they passed the annual examination they received a small government grant to supplement their salaries.[15] Monitors received a lower salary and were not examined annually, but many of those in Winchcombe went on to become pupil teachers.

After a pupil teacher had passed all her examinations she could expect her salary to rise significantly on securing a position as a certificated teacher. For example, in 1877 a newly appointed pupil teacher in Winchcombe Infants School earnt £8-10-0d per annum – twice as much as the monitress – which then increased by thirty shillings each year, while the salary of the schoolmistress was £70 plus a rent-free house to live in.[16] This was far more than could be earnt by domestic service, factory work or dressmaking, which were the most common jobs open to young women, so there was a strong financial incentive to qualify as a schoolmistress.[17]

Since a pupil teacher was responsible for the educational progress of their class from one annual inspection to the next they needed to be intelligent and capable of controlling up to fifty pupils in their charge. It is clear that some girls were identified as potential teachers when they were still at school, as the following log book extracts illustrate:[18]

> Received of Mr Plumbe the Honor certificate awarded to Mary Reeks, Ada Cummings and Lizzie Henney.

> I have appointed M Reeks, A Cummings and J Shakespeare to teach Standard I occasionally.

Three months later, however, we see another entry:

> The elder girls are, as usual, falling off – M Reeks for service, L Agg and J Shakespeare are wanted at home.

This suggests that bright girls were just as likely to end up 'going into service' or helping their mothers with domestic duties as training to become teachers. This chimes with the then popular belief that 'the home was more important than the school in the upbringing of girls'; they were also much more likely to be absent from school than their brothers to help their mothers look after younger children and assist with household chores, especially on washing day.[19]

Nepotism

One notable example of apparent nepotism in Winchcombe occurred in 1864 when, twelve months after her appointment, the schoolmistress, Margaret Malins, recorded in the log book '[My] sister, Jane Malins, entered upon her duties as pupil teacher'.[20] Since Jane previously lived with her parents in Oxfordshire Margaret had presumably recommended her to fill the pupil teacher vacancy, and the school managers were probably happy with the proposal, perhaps because it saved the expense of advertising. However, although this appears to be a clear example of family influence being brought to bear, there would have been significant consequences for Margaret if her younger sister, Jane, had proved not to be up to the job as a pupil teacher. The introduction of the Revised Code in 1862 meant that government grants to elementary schools were calculated largely on an annual examination of each pupil, except for infants, based on the three 'Rs' and conducted by one of Her Majesty's Inspectors (HMIs).[21] The continuing employment of a schoolmistress was thus dependant on the performance of the pupils in her school, which in turn meant that her pupil teachers needed to be good at their jobs; all the more reason for a schoolmistress to ensure that she had competent pupil teachers and monitresses in her charge.

Illustration II. *Mary Hawkins,*
Mistress of Gretton School (1866-75)
and Dents School (1875-82)
Photo © Janey Hawkins

Illustration III. *Thomas Hawkins,*
Master of Gretton School
(1875-1885)
Photo © Janey Hawkins

Perhaps the most notable family of teachers in Winchcombe was the Hawkins family. In 1863 Mary Hawkins was a fifth-year pupil teacher, so she may have begun her teaching career there, since the school opened in 1857 when she would have been twelve years old.[22] After attending St Marys Hall College, Cheltenham, Mary returned to Winchcombe parish in 1866 to take up post as mistress at Gretton School.[23] Nine years later, in 1875, she became the mistress at Winchcombe Girls School, the resultant vacancy at Gretton School being filled by her older brother Thomas who had previously been the Master of the Trinity Church School in Bradford on Avon, Wiltshire.[24] A mere eight days after Thomas took up post at Gretton his daughter, Bess, commenced her duties as a pupil teacher at Winchcombe Girls School under the tutelage of her aunt, Mary Hawkins. Eleven days later another of Thomas's daughters, Amy, began working as pupil teacher at Gretton School under her father's instruction. Two years afterwards, in 1877, Thomas Hawkins third daughter, Hester, took up post at Winchcombe Infants School.[25]

Did Mary Hawkins recommend her brother Thomas for the position at Gretton or did she just alert him to the impending vacancy? And was it coincidental that his three daughters, Bess, Amy and Hester Hawkins, all became pupil teachers in Winchcombe, or was that part of the deal when Thomas was offered the post at Gretton? Even if Mary Hawkins played no part in the appointment of her brother Thomas, the employment of his three daughters as pupil teachers smacks of family influence being brought to bear, particularly since the first two girls started work as pupil teachers within three weeks of Thomas taking up post as Master at Gretton School. But it might be argued that Thomas and Mary were best placed to judge the capabilities of his daughters and to compare them with any other local candidates, i.e., the older pupils in their schools. The school managers may have been happy to proceed on that basis because if one of the three sisters had performed badly in the annual examinations the managers could hold the head teacher responsible for employing his kin as a pupil teacher. Fortunately that did not happen but, as the following log book extract from 1880 indicates, the youngest of the three girls, Hester Hawkins, did not complete her indentures at the Infants School:

> The state of HM Hawkins health necessitates the removal of her name from the register of pupil teachers serving in this school.[26]

However, eleven years later Hester was an assistant mistress at a school in Bromsgrove, where her older sister Bess was the Mistress – perhaps another example of family influence being brought to bear – and ten years after that Hester was a schoolmistress in Bath, so she presumably gained her teaching certificate after leaving Winchcombe.[27]

Positive discrimination did not appear to stop when Thomas Hawkins left Gretton School in 1885 to teach at Malvern.[28] The mistress who took over from him was Sarah Lavinia Greenhalf, who two years later would become his daughter-in-law through her marriage to his son Robert, who himself was a music teacher. Sarah was from another 'teaching family' in Winchcombe – her sister Georgina began as a pupil teacher at the Girls School in 1869, her sister (or cousin), Kate, commenced working as a monitress at the Infants School in 1872, and in 1876 a cousin, Christiana, began as a monitress at the Girls School.[29] Sarah's brother, Arthur Greenhalf, was a monitor at the Boys School from 1887 but three years later his name was removed from the register of pupil teachers serving in the school.[30] However, a few months afterwards Arthur resurfaced as a pupil teacher at the National School in Brighton where the headmaster was his brother-in-law (the husband of his sister Georgina) – on the face of it another apparent example of nepotism at work, which is given some credence by the fact that Arthur was then living with his sister Georgina and her husband. Arthur went on to gain his teaching certificate in 1892 and was then a schoolmaster for over thirty-five years.[31]

Illustration IV. *The old school at Gretton*
Photo © Janey Hawkins

Gretton School provides another two examples of apparent family nepotism. The first of these began in May 1890 after Matilda Gardner, aged eighteen, a pupil teacher at Winchcombe Infants School, left after having failed her fourth year exams there the previous year.[32] The following month, June 1890, she took up post as assistant mistress at Gretton School,[33] where the schoolmistress was Annie Deans who was then courting Matilda's brother, Charles. This we know because Annie and Charles were married later that year, and since their first child, Dora, was born in March 1891 then she would have been conceived in June 1890 or shortly thereafter. Annie Deans 'strongly recommended' Matilda to the School Board, no doubt highlighting that since Matilda was almost qualified she would be an asset as a teacher at Gretton.[34] However, this appointment was not without consequences, as revealed by the report of the next HMI inspection in the following year:

> As M Gardner is not qualified for recognition as an assistant teacher under Article 50, a deduction has been necessary under Article 108, on account of the insufficiency of the staff during the past year.[35]

This deduction amounted to £11, one fifth of the annual grant, thereby reducing it from £57 to £46. So, Annie's desire to help Matilda, her future sister-in-law, proved very costly for the school, and demonstrates that not all family collaborations were successful.[36]

The second example of family nepotism at Gretton School occurred twenty years later when Conrad New was the schoolmaster, and the log book records that in 1909 Irene Minett, aged fourteen, started work as a monitress. Five years later, in 1914, her elder sister Madeline, aged twenty-six, took up post there as a supplementary teacher, and in 1918 a third sister, Audrey, aged sixteen, began working there as a pupil teacher. In itself three sisters working at the same school in less than a ten year span might seem unusual, but perhaps less so if it were known that they were the nieces of the schoolmaster's wife, Mrs Bertha New (née Minett), herself a former monitress at the school.[37]

Many of the above examples suggest that some school managers may have taken the view that if a pupil teacher proved to be successful and their younger siblings or cousins were academically inclined then they would also make good teachers. When Eva Pullom began as a pupil teacher at the Infants School in 1869 she would have been aware that her first cousin, Catherine Seabright, was then a second year pupil teacher at the Girls School. The following year their second cousin, Emma Seabright, joined Eva at the Infants School as a monitress, before transferring to Gretton School the year after as a pupil teacher. Eighteen months later, in 1873, Emma's sister, Mary, started work as a monitress at the Infants School, and nine years after that their cousin Edith Cummings became a monitress there.[38] Was it a coincidence that these five cousins all became pupil teachers or were Catherine Seabright's kin deliberately selected to follow in her footsteps, perhaps because she was a promising pupil teacher who went on to win a first class scholarship to Cheltenham College?[39]

Similar observations could be made about the Lishman family. Sarah Agnes Lishman began as a monitress at the Infants School in 1875, her sister Annie was described as a teacher in 1881 (possibly employed at a local private school), and another sister, Lavinia, began as a monitress at the Girls School in 1884.[40] Their niece, Maude Lishman, was a monitress at the Girls School from 1898, and another niece, Gladys Haslum (whose father, Walter Haslum, was a local music teacher), began as a monitress there in 1908.[41]

Another 'teaching family' was the Kings. In 1873 Fanny Kings began teaching at the Girls School and twelve months later she transferred to the Infants School.[42] Her sister, Harriett, followed a similar path, starting as a monitress at the Girls School in 1875 and transferring to the Infants School two years later, and another sister, Annie began as a pupil teacher at the Girls School in March 1875.[43] In 1876 their twelve-year-old brother, Thomas, was described as having ceased teaching at the Boys School, so he had presumably not been there for very long.[44] Finally, in 1882 their twelve-year-old sister, Lizzie, was described in the Girls School log book as:

A promising pupil [but] 'is wanted at home' and will therefore fail to make the required attendance[45]

which suggests that domestic responsibilities may have prevented Lizzie Kings from following in the footsteps of her four siblings.

Charles Martin and his three siblings (Edith, John and Frank) started work as monitors or pupil teachers between 1876 and 1897,[46] and likewise four members of the

Hall family (Emma née Child, Caroline, Conrad and Dora) taught in Winchcombe between 1861 and 1899.[47] In addition, there were three other occasions when three family members were teachers (from the Pearson, Furley and Harvey families), plus a further eight cases where two family members were teachers (Malins, Taylor, Medcraft, Wood, Smith, Healey, Beach and Minett). In fact, out of the 139 pupil teachers and monitors who worked in Winchcombe's four National schools between 1860 and 1910, forty-eight of them became fully fledged teachers – whether certificated or uncertificated – and of those forty-eight, over two thirds of them were related to another local teacher.

Illustration V. *Dents school and the school house, which opened in 1868 - in 1876 it became the Girls School. Source unknown*

The existence of these 'teaching families' in Winchcombe parish suggests that school managers may have preferred to appoint family members, and this could have been a way of ensuring reliability and guaranteeing that the pupil teachers would commit to the hard work necessary to meet the increasingly exacting standards of the HMI.

TABLE I

Pupil teachers and monitors in Winchcombe's National/Board schools, 1860-1910

	PTs and monitors	PTs and monitors who became teachers	Teachers related to local PTs/monitors	Teachers not related to local PTs/monitors
Infants School	35	16	12 (75%)	4
Girls School	48	18	12 (66%)	6
Boys School	23	9	7 (77%)	2
Gretton School	33	5	2 (40%)	3
TOTALS	**139**	**48**	**33 (68%)**	**15**

Other influences

The owners of Sudeley Castle, situated one mile from Winchcombe town, exerted a considerable influence over the schools in Winchcombe. John Coucher Dent had inherited the castle from his uncle (John Dent, a glove manufacturer, originally from Worcester), and he lived there from 1855 as de facto lord of the manor. He was a barrister and magistrate and subsequently became lord lieutenant of Gloucestershire, but he still found time to take an interest in local affairs. As well as founding Dents School in 1868, funded largely by his uncle's legacy, he was the first chairman of the Winchcombe School Board, and his death in 1885 was marked by the granting of a half holiday at Dents School and the Infants School.[48] His wife, Emma, took a close interest in Winchcombe's National Schools – especially Dents School, no doubt because of the family connection – and she made frequent visits to all four of them. Typically the log books record that she would call in to one of the schools with some friends and ask to hear the children sing or read, or she would take materials for needlework. She often officiated at the annual prize-giving ceremonies for good attendance, and hosted treats each year for pupils, sometimes as many as 550 children at a time and, separately, for the staff of all four schools.[49]

Closer examination of the log books indicates that Emma Dent had a significant influence on the issues that affected local schools, including staffing matters. For example, shortly after she found out that the schoolmistress at Dents School, Miss Emma Child, was heavily pregnant in late 1862 Emma Dent recorded in her diary, whilst staying in London, that:

> I went to the great Educational Department in the Sanctuary to enquire about a school mistress – a list is kept there of all wanting situations.[50]

The Sanctuary was the location of the Department of Education, and referring to its list of unemployed or newly qualified teachers was an easy way to find someone who would be able to quickly take over from Miss Child.

Furthermore, in 1864 when the schoolmistress was ill Emma Dent 'took the French classes and Geography', and in 1879 the schoolmistress recorded in the log book that 'the school was reopened by the pupil teachers as Mrs Dent kindly allowed me another weeks holiday'.[51]

Two more extracts in the Dents School log book, from 1880 and 1883, are quite telling:

> Mrs Dent appointed Miss B Hawkins as assistant mistress until the end of July,

> I find E Monk incapable to perform her duties as candidate on probation – have acquainted Mrs Dent with the fact. It is decided for her to leave shortly and that Sarah Johnson should take her place,

which suggests that in addition to being one of the school managers at Dents School, Emma Dent was also the first point of contact for its schoolmistress, and had the authority to appoint members of staff and deal with teachers who did not reach the required standards.[52]

The following extracts from the Gretton School logbook in the 1860s tell a broadly similar story:[53]

> Heard from Mrs Dent that a monitor is coming on trial, to commence duty tomorrow.
> Visit from Mrs Dent this morning, promised to get another monitor in Alice's stead.
> E Richardson, a paid monitor provided by Mrs Dent, came on trial today.
> Gave the children a holiday by Mrs Dent's permission.

In addition, in 1874 the Chairman of the Gretton School Board consulted with Emma Dent when Miss Hawkins, the schoolmistress, resigned because her request for an increase in salary to £70 per annum was refused. Emma then suggested ways in which the mistress's services might be retained, 'and offered, with the view of retaining the services of Miss Hawkins, to guarantee the amount of salary above mentioned for one year'.[54]

When the government issued orders for the compulsory formation of a school board in the parish of Winchcombe, Emma's husband, John, stood for election as a member, and he served as Chairman of the Winchcombe United School Board from its inception in 1875 until his death ten years later.[55] Although it was not usual for women to serve on public bodies, Emma was then asked to take her husband's place on the school board, although when she stood down at the next triennial election in 1887 she confided in her diary that 'she was not sorry to think it was all over'.[56] Although it was customary for the gentry to take an interest in the schools in their locality, it was less usual for them to be involved in day to day issues, but Emma Dent was well-respected and her involvement no doubt helped to 'make acceptable the patriarchal social system'.[57] She had no children of her own, but her close involvement with the local elementary schools extended far beyond the activities that were usual for a lady of the manor in the local community. Her particular devotion to Dents School was demonstrated after her death in 1900, when it was revealed that in her will she bequeathed £1 to each of its teachers, and 2s 6d and a black frock to each child there.[58]

There is further evidence that head teachers in Winchcombe were sometimes able to influence pupil teacher appointments because in 1882 Elizabeth Freeman, mistress of the Infants School, was asked to decide which of two monitresses should be taken on as a pupil teacher. Of the girls under consideration, Lizzie Henney had been awarded an Honour Certificate four years earlier when she was a pupil at the Girls School and had subsequently 'left for private school after passing five standards'. The other candidate was Georgina Newman, who had been listed in the 1881 census as a general servant, working for and living at the home of the mistress, Elizabeth Freeman, although she was also a monitress at the Infants School. Despite Lizzie Henney's excellent academic record Miss Freeman chose Georgina Newman to fill the vacant post and she subsequently went on to college and had a successful teaching career. Presumably as a consequence of not securing the pupil teacher position the log book entry for the following day read 'L Henney expressed a wish to resign, not liking the work'.[59]

Another example of a schoolmistress being involved in the selection process for new teachers occurred in 1896 when Emily Catton, the newly-appointed mistress of the

Infants School, was asked by the managers to choose between four applicants to fill the vacancy of monitress. In order to best assess their capabilities each of the girls spent a week teaching on trial at the school so the mistress could determine which was 'the most adapted for a teacher'. In the end Lily Beach, the first girl to attend, was subsequently appointed, but two of the other three girls later became monitresses in other local schools.[60]

It would appear, however, that head teachers did not always get the pupil teachers they wanted. As a scholar at the Boys School, George Brewer was in trouble for stone throwing, his attendance was poor and the Master described him as 'a bad swearing boy'. Despite that, three years later in 1885 at the age of fifteen, and by then a pupil at the local grammar school, he began work as a pupil teacher at the Boys School. This appointment was made by the School Board and against the wishes of the Master, Mr Kemp, who presumably remembered Brewer as a badly behaved former pupil.[61] Other pupil teachers at the Boys School also failed to meet the Master's approval. For example, Conrad Hall was sent home for misconduct during lessons, was unpunctual, his work was described as very poor, and he was described by the Master on separate occasions as a fool and extremely dense.[62] Surely the Master would not have selected such a poor prospect to be a pupil teacher – or did his mother, Emma Hall (née Child), a former local schoolmistress, exert some influence with the School Board. [Nevertheless, despite being such apparently poor prospects both George Brewer and Conrad Hall became schoolmasters!][63]

Changing practices

The extent of apparent nepotism in the appointment of young teachers in Winchcombe parish in the mid to late nineteenth century prompts the question as to whether the school managers may have actually preferred to appoint relatives, because in many cases the newcomers would have come with good references from existing trusted staff. Furthermore, this chimes with the contemporaneous practice of artisan fathers often finding jobs for their sons through family connections, which applied in many working class occupations including agriculture.[64] But the nepotism in the appointment of young teachers in Winchcombe did not go unnoticed, because just prior to the school board election of 1902 one of the candidates pointed out that three of the existing board members (Conrad Hall senior, John Martin and George Troughton), who were standing for re-election, 'had tried or were trying to obtain situations for their children under the board' and would therefore not be unbiased on issues such as cutting salaries. Nonetheless, Hall, Martin and Troughton were duly re-elected as board members.[65]

By the late nineteenth century teaching was not yet a profession but it had made huge strides compared with earlier in the century when a village school teacher was as likely to be an impoverished working man 'unfit for manual labour' as a man of letters; indeed one teacher in early nineteenth-century Winchcombe was 'named Turner, an agricultural labourer with a talent for elementary arithmetic but innocent of grammar and composition'.[66]

Certainly by the early twentieth century any nepotism in the appointment of young teachers appeared to have dwindled in Winchcombe because in 1902 the School Board decided to advertise for two pupil teachers for the Boys School,[67] but the following extracts from the log book three years later show the frustration of the Master, Alfred Morris, over his non-involvement in the selection process:-

> Hubert Woodward has been appointed monitor during the holidays with a view to his ultimate engagement as a pupil teacher. Although the Official Correspondent has spoken to me about the appointment some time ago the managers have not consulted me in any way about the matter or the fitness of the candidate for the post.[68]

This indicates that it was still the practice for pupil teachers at the Boys School to be recruited by the School Board with little input from the Master. The likely reason for this had been alluded to in an earlier board meeting when two of the members stated that 'the present master was unable to keep pupil teachers whereas previous masters always had'.[69] This is borne out by an analysis of relevant entries in the log book, which shows that in the eight years after Alfred Morris's appointment as Master in 1894 only one of his pupil teachers had become qualified, whereas in the last eight years of the tenure of his predecessor, George Kemp (1876 to 1894), at least five of his pupil teachers became fully fledged teachers.

Another extract from the Boys School log book, later in 1905, shows Alfred Morris's continuing dissatisfaction with the recruitment process:-

> I also called attention to the desirability of interviewing candidates for assistantships in the school. These appointments are made in a most unsatisfactory manner.

However by 1907, by which time school boards had been superceded by local education authorities under the Education Act of 1904, things were changing and the Master of the Boys School wrote in his log book:-

> I have recommended the appointment of Mr C P Lewis as Certified Assistant vice Mr Palmer,

and the next year he recorded:-

> On Thursday I had an interview with Mr H Bishop of Ross who is an applicant for the post vacated by Mr Lord,

followed a few weeks later by:-

> In accordance with instructions received from the Education Committee I had an interview with Mr Chas Edward Cooper, a candidate for the vacant post of uncertified teacher in this school,

which shows that although head teachers still had a formal role to play in the recruitment of staff, the control of appointments was by then firmly in the hands of the local authority.[70] These later log book extracts illustrate the closing stages of the drawn-out transition from a recruitment process often based on family connections to a 'competence-based' job market for teachers.

Conclusion

From the foregoing it is clear that in the second half of the nineteenth century nepotism was sometimes brought to bear in the appointment of pupil teachers and monitors in Winchcombe's national and board schools, with head teachers at times influencing the appointment of members of their own family; at Gretton School this appears to have continued until well into the twentieth century. Whilst some head teachers had a say in the selection of staff, at the Boys School it appears that the School Board were slower to involve the head teacher in recruitment issues and may not have done so until after 1904, when Winchcombe's four elementary schools came under the control of Gloucestershire County Council, which subsequently issued recruitment guidelines to school managers.

The high number of pupil teachers and monitors in Winchcombe between 1860 and 1910 who were related to each other suggests that school managers often appointed a relation of a pupil teacher or monitor because they presumed – or hoped – they would be 'cut from the same cloth'. Although this may have been a sign that head teachers were simply trying to ensure that they had capable staff, it was in fact a good indicator of likely success because, overall, more than two thirds of those who became fully fledged teachers were related to another local teacher. Furthermore, this practice may simply have reflected the usual way in which youngsters then obtained jobs, i.e., through family connections.

The one person who arguably had the greatest influence over the appointment of pupil teachers in the four Winchcombe national and board schools was Emma Dent of Sudeley Castle, and her involvement and support of the head teachers and their staff covered a span of well over thirty years, extending into the 1890s, which underlines the importance of local hierarchical social connections in the provision of teachers before the days of local authorities.

Notes

1. Horn, P. *Education in Rural England 1800-1914*. Gill and Macmillan, 1978, p. 121.
2. Cruikshank, M. *Church and State in English Education*. Macmillan and Co, 1963, p. 8; Kay-Shuttleworth, J. *Four Periods of Public Education*. London, 1862, pp. 483-5.
3. Light, A. *Common People: The History of an English family*. Fig Tree, 2014, p. 227.
4. Horn. 1978, pp. 66-68 and 76.
5. Gloucestershire Archives (GA), S368/1/3/1-2; Winchcombe Girls School log book (GSL), 19 March 1866, 1 April 1867 and 2 March 1868.
6. *Gloucester Journal*, 27 April 1878; *Gloucester Citizen*, 27 September 1879.
7. GA, P368/1 OV7/3, Census return, 1831; Donaldson, D. *Winchcombe: a history of the Cotswold borough*. Wychwood Press, 2001, p. 69, 125; Victoria County History. *A History of the County of Gloucester*. Vol 2, 1907, Population tables, p. 182.
8. *Cheltenham Chronicle*, 13 January 1857.
9. Emma Dent's diaries: 12 January 1862 and 1 January 1868.
10. GA, S368/1/3/1, GSL, 31 March 1876.
11. GA, P368/1 SC2/5, Gretton School accounts, vouchers and correspondence.
12. *Evesham Journal*, 27 February 1875; *Cheltenham Chronicle*, 12 May 1900.

13. Wright S. 'Teachers, family and community in the urban elementary school: evidence from English school log books c.1880 - 1918', *History of Education Journal*, 2012, 41, No. 2, p. 159.

14. Horn. 1978, p. 76; Evans, K. *The development and structure of the English Educational System.* Hodder and Stoughton, 1975, p. 31.

15. Horn. 1978, p. 74.

16. *Gloucestershire Echo*, 26 April 1884.

17. Horn, P. *The Victorian and Edwardian Schoolchild.* Alan Sutton Publishing Ltd, 1989, p. 167.

18. GA, S368/1/3/1, GSL, 20 May 1878, 19 April 1880 and 12 July 1880.

19. Dyhouse, C. *Girls growing up in late Victorian and Edwardian England.* Routledge & Kegan Paul, 1981, p. 101-2.

20. GA, S368/1/3/1, GSL, 18 January 1864.

21. Horn. 1978, p. 125.

22. GA, S368/1/3/1-2, GSL, 23 October 1863.

23. Thomas Hawkins diary, December 1865; GA, S368/4/1//1, Gretton School log book (GrSL), 11 January 1866.

24. GA, S368/1/3/1, GSL, 5 April 1875; GA, S368/4/1//1, GrSL, 23 April 1875; Thomas Hawkins diaries, 1859 to 1875.

25. GA, S368/4/1//1, GrSL, 28 June 1875; GA, S368/1/3/1, GSL, 6 July 1875; GA, S368/1/1/1, Winchcombe Infants School log book (ISL), 4 May 1877; Thomas Hawkins diaries, 17 July 1875 and March 1877.

26. GA, S368/1/1/1, ISL, 23 April 1880.

27. Census Return 1891, RG12/2345, f.59 p.4; Census Return 1901, RG13/2343, f.105 p.17.

28. Thomas Hawkins diary, May 1887.

29. GA, S368/1/1/1, ISL, 7 October 1872; GA, S368/1/3/1, GSL, 3 February 1869 and 29 February 1876.

30. *Gloucestershire Echo*, 31 May 1887; GA/S368/1/2/1, Winchcombe Boys School log book (BSL), 16 June 1890.

31. Census Return 1891, RG12/822, f.98 p.33; www.findmypast.co.uk, Teachers Registration Council, Registration No. 53396, accessed 2014.

32. GA, S368/1/1/1, ISL, 15 June 1889 and 23 May 1890.

33. GA, S368/4/1//1, GrSL, 9 June 1890.

34. *Cheltenham Chronicle*, 3 May 1890.

35. GA, S368/4/1//1, GrSL, 15 May 1891.

36. *Gloucestershire Chronicle*, 16 May 1891.

37. GA, S368/4/1/1-3, GrSL, 11 October 1909, 19 January 1914, 19 August 1918 and 3 June 1884.

38. GA, S368/1/3/1, GSL, 6 May 1869; GA, S368/4/1//1, GrSL, 15 December 1871; GA, S368/1/1/1, ISL, 12 February 1869, 20 June 1870, 12 May 1873 and 23 June 1882.

39. GA, S368/1/3/1, GSL, 3 February 1872.

40. GA, S368/1/1/1, ISL, 11 June 1875; Census Return 1881, RG11/2564, f.157 p.4; GA, S368/1/3/1, GSL, 28 April 1884.

41. White, R. 'Winchcombe's Musical Maestro', *Winchcombe History Journal*, 2018, pp. 43-48; GA, S368/1/3/2, GSL, 26 July 1898 and 2 October 1908.

42. GA, S368/1/3/1, GSL, 5 April 1873; GA, S368/1/1/1, ISL, 15 June 1874.

43. GA, S368/1/1/1, ISL, 4 May 1877; GA, S368/1/3/1, GSL, 26 February and 19 March 1875.

44. GA/S368/1/2/1, BSL, 14 April 1876.

45. GA, S368/1/3/1, GSL, 13 Apr 1882.

46.   GA/S368/1/2/1, BSL, 18 February 1876, 19 April 1880 and 27 April 1896; GA, S368/1/3/1, GSL, 24 March 1876.

47.   Census Return 1861, RG9/1792, f.11 p.16; GA, S368/1/3/1, GSL, 23 October 1863 and 10 January 1896; GA/S368/1/2/1, BSL, 8 January 1877.

48.   *Bath Chronicle and Weekly Gazette*, 25 October 1855; *Evesham Journal*, 27 February 1875; *Gloucester Journal*, 27 March 1875; GA, S368/1/3/1, GSL, 30 March 1885; GA, S368/1/1/1, ISL, 2 April 1885.

49.   GA, S368/1/3/1, GSL, 27 July 1866.

50.   Emma Dent's diary, 5 November 1862.

51.   GA, S368/1/3/1, GSL, 2 December 1864 and 8 September 1879.

52.   GA, S368/1/3/1, GSL, 4 June 1880, 29 October to 2 November 1883, and 1 August 1872.

53.   GA, S368/4/1//1, GrSL, 13 June 1864, 20 September 1864, 2 July 1866 and 9 October 1866.

54.   GA, P368/1 SC2/3, Gretton School Managers' minutes, 1864-1874.

55.   *Gloucester Journal*, 6 February 1875; *Cheltenham Mercury*, 04 April 1885.

56.   *Cheltenham Chronicle*, 28 April 1885 and 26 February 1887; Dent's diary: 29 January 1887.

57.   Horn, P. *Ladies of the manor: Wives and Daughters in Country-house Society 1830-1918*. Alan Sutton Publishing Ltd, 1991, p. 113.

58.   *Bath Chronicle and Weekly Gazette*, 20 September 1900.

59.   www.findmypast.co.uk, Teachers Registration Council, Registration No. 5772, accessed 2014; GA, S368/1/1/1, ISL, 29 September 1880, 15 and 16 June 1882.

60.   GA, S368/1/1/2, ISL, 31 August 1896.

61.   GA, BSL, GA/S368/1/2/1, 4 July 1882; Winchcombe Folk and Police Museum, WIXFP 2010.1060, Chandos School Trustees minute book, 17 January 1887; *Cheltenham Chronicle*, 3 February 1885.

62.   GA, BSL, GA/S368/1/2/1, 26 October 1877, 10 December 1877 and 19 March 1878.

63.   Census Return 1891, RG12/4116, f.73 p.23, and RG12/4594, f.63 p.2.

64.   Vincent, D. *Bread, knowledge & freedom: a study of nineteenth century working class autobiography*. Methuen, 1981, p. 66.

65.   *Cheltenham Chronicle*, 22 February 1902.

66.   Horn. 1978, p. 15; *Evesham Journal*, 27 February 1875.

67.   *Cheltenham Chronicle*, 21 June 1902.

68.   GA, BSL, GA/S368/1/2/1, 15 September 1905.

69.   *Cheltenham Chronicle*, 13 December 1902.

70.   GA, BSL, GA/S368/1/2/1, 8 December 1905, 18 December 1907, 18 September 1908 and 28 October 1908; Sellman, R. 'The Country School', in *The Victorian Countryside*. Mingay, GE (ed.), Vol. 2, Routledge & Kegan Paul, 1981, p. 551.

Bibliography

Primary Sources
Gloucestershire Archives (GA)
        Gretton School log books (1864-1934), GA/S368/4/1/1-3
        Gretton School Accounts (1864-74), GA/P368/1 SC2/5
        Gretton School, Managers Minute Book (1864-74), GA/P368/1 SC2/3
        Population enquiry (1831), GA/P368/1 OV7/3
        Winchcombe Boys School log book (1874-1913), GA/S368/1/2/1
        Winchcombe Girls School log books (1863-1913), GA/S368/1/3/1-2
        Winchcombe Infants School log books (1868-1934), GA/S368/1/1/1-3

Other primary sources
  Bath Chronicle and Weekly Gazette
  Census Enumerators Books (1841 – 1911)
  Chandos School Trustees minute book (1831-2004), Winchcombe Folk
    and Police Museum
  Cheltenham Chronicle
  Emma Dent's diaries
  Evesham Journal
  Gloucester Citizen
  Gloucester Journal
  Gloucestershire Chronicle
  Gloucestershire Echo
  Thomas Hawkins diaries

Secondary Sources
Cruikshank, M. *Church and State in English Education*. Macmillan and Co. 1963.
Donaldson, D. *Winchcombe: a history of the Cotswold borough*. Wychwood Press, 2001.
Dyhouse, C. *Girls growing up in late Victorian and Edwardian England*. Routledge & Kegan
    Paul, 1981.
Evans, K. *The development and structure of the English Educational System*. Hodder and
    Stoughton, 1975.
Horn, P. *Education in Rural England 1800-1914*. Gill and Macmillan, 1978.
Horn, P. *The Victorian and Edwardian Schoolchild*. Alan Sutton Publishing Ltd, 1989.
Horn, P. *Ladies of the manor: Wives and Daughters in Country-house Society 1830-1918*. Alan
    Sutton Publishing Ltd, 1991.
Kay-Shuttleworth, J. *Four Periods of Public Education*. London, 1862.
Light, A. *Common People: The history of an English family*. Fig Tree, 2014.
Sellman, R. The Country School, in *The Victorian Countryside,* Mingay, GE, Vol. 2, Routledge &
    Kegan Paul, 1981.
Victoria County History. *A History of the County of Gloucester*. Vol. 2, 1907.
Vincent, D. *Bread, knowledge and freedom: a study of 19th century working class autobiography*.
    Methuen, 1981.
White, R. 'Winchcombe's Musical Maestro'. *Winchcombe History Journal*, 2018, pp. 43-48.
Wright, S. 'Teachers, family and community in the urban elementary school: evidence from
    English school log books c.1880–1918'. *History of Education Journal*, 2012, Vol. 41, No.
    2, pp 155-173.

Digital Resources
www.findmypast.co.uk. Teachers Registration Council Registers (1914-18).
www.genuki.org.uk. The Hawkins Family and their contribution to Education in 19th Century
    Winchcombe, formerly published at www.genuki.org.uk and accessed July 2014 but
    currently unavailable.

Acknowledgements

I would like to thank Lady Ashcombe of Sudeley Castle, for allowing me to use
information from Emma Dent's diaries, and Jean Bray, retired archivist at Sudeley Castle,

for her advice and guidance. In addition, I am grateful to Janey Hawkins (1929-2015) for letting me use information contained within the diaries of Thomas Hawkins, who was Master of Gretton School from 1875 to 1885, and from her website 'The Hawkins Family and their contribution to Education in 19th Century Winchcombe'.

Biographical Note

Rob White is a former DA301 student (Open University) whose subsequent Masters Research degree was about local government in late eighteenth- and early nineteenth-century Winchcombe. His interest in Winchcombe emanated from a search for his wife's family origins, and he was a founder member of a local history society there (the Gloucester Street History Group) in 1990, which is still actively involved in researching and writing about various aspects of the history of Winchcombe.

"IT'S NOT ALWAYS MY FAULT":
A STUDY OF NON-ATTENDANCE AT TWO BRADFORD SCHOOLS
IN THE NORTH OF ENGLAND, 1880 TO 1900
Ray Greenhough

School Log Books contain a wealth of information which can be used to survey attendance levels at Victorian schools. Pupil attendance is a topic regularly visited by social historians. Their research has suggested that the causes of absence were usually the fault of the child or its parents. Previous studies have neglected to consider whether the school itself was a contributory factor in poor attendance levels. This article, while discussing the broad issues that historians, such as Susannah Wright, have raised in the past, takes as its focus the schools themselves and their physical environment.[1] It will argue that the school logbook can provide new historical insights into the patterns of school absence, which were shaped by the modus operandi *of a reforming educational system and associated socio-economic trends. This article will in this refined case-study analysis use the logbooks from the Chapel Green Board and Church schools based at Bradford in the North of England from 1880 to 1900.*

Introduction

Prior to the Elementary Education Act, the Anglican Church was the main provider of education in England.[2] Geoffrey Best first suggested in the 1950s that the Church of England's ethos was to exert 'its influence both in life and thought' across Victorian society.[3] Furthermore they considered themselves as 'part of the constitution itself' by virtue of their Established status and thus 'took it for granted that it must retain certain marks of superiority and privilege' in education.[4] After the new first Elementary Education Act was passed, all Church of England schools were henceforth in competition with newly constituted non-denominational schools known as Board Schools.[5] These were supposed to be non-denominational, supplemented by a national education grant paid for from taxation. Naturally most non-conformists preferred to send their offspring to attend these new Board Schools.[6] It was thus the culmination of a century of educational debates[7] in which the sector diversified by having voluntary schools (generally Catholic or Church of England) alongside Board Schools (elected triennially in the boroughs by the burgesses and in parishes by ratepayers, and were given the power to issue a precept on the rating authority to be paid out of the local rate). The question of who was and who was not entitled to claim educational grants from taxation was however to prove troublesome.[8] It was only resolved at the general election of 1874 when the Liberal Party won a victory on a ticket of political compromise.[9] Essentially, Gladstone allowed voluntary schools and their non-denominational counterparts to claim grants provided they improved attendance rates and teaching standards, whilst reducing absences subject to a system of inspection.[10] Being far more experienced in education, the Church of England schools were thus better placed to resolve any future potential attendance issues to obtain their grants, whereas their competitors, the new Board Schools, were inexperienced players in the field of education and could only fill the gaps left by the established Church. As a result, it took the latter a long time to overcome their attendance problems and to catch up with their Anglican counterparts – the central focus of this article.

There has been a large volume of literature published about the reasons why children did not attend school and school logbooks have made an important contribution to these debates. In what follows, keynote debates in the historiography are synthesised to place this School Logbooks booklet into context. In the main, studies have concentrated on the sociological causes of why either parent or child chose whether or not to attend school. Parental choice, local attitudes, family economy, apathy and expectations are often quoted as reasons for poor attendances. John Hurt, for example, argues that parental choice before the 1870 Elementary Education Act was 'the parent choosing if and how much education their child should receive'.[11] After the legislation was passed he views their choices as being eroded by the State's increasing intervention with 'patterns of life by coming between parent and child and imposing new patterns of behaviour'.[12] No longer 'could parents take for granted the services of children in the home and their contribution to the family budget'.[13] A further Elementary Education Act[14] now made attendance compulsory; however, despite the State's attempts to exert pressure in order to get children into the school room, absenteeism remained high. This may have been motivated by two socio-economic causes: firstly, compulsory education was not free; and secondly, this financial fact exacerbated the loss of family income when children stopped working. Some relief in respect of this financial problem occurred in 1891 when education finally became a right without cost under a further Elementary Education Act in 1891.[15] The Bradford School Board, however, concluded that in their opinion 'the abolition of school fees has exercised no influence upon the attendance of children'.[16]

ACO Ellis argues instead that parental attitudes towards education resulted in 'an element of active hostility towards education for which there were several obvious reasons'.[17] Hurt and FML Thompson likewise both claim that this antipathy revolved around the supposed attempt to exert a high degree of social control over all Victorian children per se. Various studies have claimed that the socialising function of the school 'involved the regulation of thoughts and habits of the children'.[18] Thompson thus shows that there is evidence of schools providing the 'objective needs of society to guide, restrain and control its members so they obey accepted conventions in thoughts and behaviour'.[19] Steven Humphries supports this view saying that this function resulted in 'the bureaucratic organisation of school and work which reduced and regulated leisure time' and thus formed an important reason for absenteeism.[20] Truancy, he goes on to state, on the one hand 'was an expression of the conflict between working class recreation demands which were based on personal impulses and community traditions'.[21] On the other hand, parental apathy was recognised by many contemporary commentators and subsequent historians of education as a major factor of absenteeism. Hurt implies that parents hence viewed the need for education through 'themselves and their own social status as the standard up to which they proposed to educate their offspring'.[22] Nonetheless, Jonathan Rose questions the validity of this argument and instead advocates that those who proposed these theories had drawn their 'information from official sources' and that the 'parents rarely spoke in them'.[23] Self-evidently then there were conflicting ideas and attitudes in circulation some of which were substantiated by parents' actions, and others which reflected broader attitudes by those in authority in what remains a vibrant area of the history of education in community settings.

This article takes then as its main focus of new research the pivotal period between 1880 and 1900 when education was formalised by the State and parents were expected to conform to a school day. Although this could be considered a short time frame, in fact it was a fascinating twenty years during which education became free and established Church of England Schools where starting to be phased out. Technically, with the advent of compulsory and free education, attendance rates should have been at their highest. Yet, comparing and contrasting two different school boards in Bradford (one Anglican and the other non-denominational) that have a rich archive base can facilitate a new evaluation of the root causes of school absence in what was two competing types of educational provision. In particular, school logbooks will be used to examine contemporary comments on pupil absence, staffing, school environment and annual inspection reports. This study will thus examine what part, if any, the schools themselves played in poor pupil attendance statistics since their role has often been neglected in educational studies. Four broad themes will be explored: firstly, this paper examines why school logbooks were introduced. Secondly, it is necessary to assess whether the schools in this study had an attendance 'problem', or not; how it was defined on location; before comparing rates of attendance with the status quo elsewhere as exemplified by local and national attendance trends. Thirdly, the chosen refined case-study analysis examines ways in which staffing and school environment encouraged or discouraged attendance. Finally, the new research asks whether annual inspector reports can support or enhance our understanding of the historical prism of school attendance picture and its broader meaning for educational standards in the late-Victorian era.

School Logbooks

In the late 1850s high levels of government spending on healthcare became an issue that needed addressing at a regional and local level. The Treasury also sought ways of reducing their commitment to funding future educational reform. In order to find ways in which this might be done from taxation, a Royal Commission was set up. Gordon Roderick and Michael Stephens point out that the extent of attendance problems that were reported in the 1862 report saw 'the average attendance was only 76% and that one-third of pupils attended for less than 100 days'.[24] Despite such poor attendance, the Commission went on to reject compulsory schooling (on a national scale) as 'neither desirable nor attainable'.[25] Instead, they recommended linking educational grants that supported schools to attendance levels and required teachers to reach minimum standards in the three Rs. The 1862 Revised Code of Education accepted the Royal Commission's views and made educational grants funded from taxation conditional. This became known as 'payment by results'. This Act triggered however considerable opposition from the Church of England who feared that it would reduce their financial ability to run schools and therefore reduce their religious influence. As a Christian nation with an established Anglican Church as its official religion they felt that they needed to teach their ethos from a young age and that the 'schools were the way to maintain it'.[26] To provide a method of audit for the government on school life, the Act also introduced school logbooks. The Church of England equally opposed this bureaucratic change, fearing that the comments in logbooks could be used against schools. One cleric was reported in the *Derby Mercury* newspaper as saying 'he was suspicious of the logbook as it was a regular spy on the whole proceedings of a school'.[27] He also feared that 'an inspector might refuse' to allow the annual education grant, having objected to 'the entries made in the logbook'.[28] To examine this controversy in closer detail it is necessary to focus on our refined case-study analysis, outlined in the introduction to this article.

The Chapel Green Board School at Bradford opened its doors to new pupils in 1877 as the result of a petition by local residents who cited overcrowding as the main reason for needing further educational provision. Why the alternative Church of England school did not simply expand its provision to avoid losing its pre-eminence is not clear, but the balance of the evidence appears to suggest that this may have been due to inadequate finances. An underlying factor appears to have been that the local Congregational Church wanted to rent space out in the Chapel Green Congregational Church, which also then accommodated the Chapel Green Board School. The school hence had two departments, mixed and infants, and occupied the ground floor, which comprised four small classrooms, a hall and a separate Infants' area. Altogether, it provided accommodation for 304 children on a daily basis. In November 1886, the Board school left the Congregational Church building and moved to a purpose-built school, changing its name to Thornton Lane Board School. In this new establishment, its structure remained unchanged, although it almost tripled in size to 810 children.

By way of comparison, Chapel Green Church School had opened slightly earlier in 1867 and used the upper floor of the Congregational Church building, which consisted of one large room. It was somewhat smaller and provided space for just 210 children in the same format as described above. Both schools where located in the village of Little Horton, which sits within the boundary of Bradford in West Yorkshire, and each had the same catchment area for their pupils. The socio-economic status of the schools' catchment area was thus one of deprivation, although it was not the worst in Bradford. The main employment opportunities within the community were in textile manufacturing and mining. Nevertheless, the makeshift economies of people in the area were fragile, and parents that sent their children to the school often lived on the critical threshold of relative to absolute poverty – factors that were to influence attendance levels at school, to which we now turn in more detail.

Attendance Levels

Figure 1 (below) sets out the national legislative framework for changing education standards, demarks when exactly Bradford adopted the new changes, and identifies the actual schools that took in pupils locally from 1880 to 1900 (the timeline of this study's focus). The first question that arises be must to ask whether Bradford, and the schools cited above, had an attendance problem. To assess this, it is necessary to examine the official Bradford attendance rates and compare them against the two schools being explored. It should be noted that the official attendance data that Bradford published was based on a calculation of *average results* from the daily data recorded by Board and Church of England schools. This article will hence seek to contextualise the two types of educational provision on that basis. It does so by setting out the quantitative data collected against that of the Bradford average and hence highlights any key differences in patterns of attendance behaviour. Secondly, by investigating Bradford attendance data against national data we can examine whether Bradford had a general attendance 'problem' or if the 'problem' was more localised in specific schools. The main source for Bradford attendance data is the Bradford School Board Triennial Reports, which are rich in information and statistics. The voices of the various committees can be heard through the section that each school completed within the overall report and this creates a sense of what was occurring within the locality. Thirdly, it is important to examine exactly how the Chapel Green Board and Church school attendance rates compared against the Bradford averages. In this way, the overall aim is to provide a performance indicator against which all the Bradford school attendance rates – established and non-denominational schools – can be evaluated.

National Educational Framework	Bradford's timeline for the adoption of the new measures	Church of England Schools	Non-Denominational Boards
Royal Commission Report (1862)		Chapel Green Church School opened 1867	
Education Act (1870)			Chapel Green Board School opened 1877
Education Act (1880)			
Education Act (1891)	↓	↓	↓

Fig. 1. Schools Provision in Bradford from 1870-1900

Figure 2 complements Fig. 1 by showing that between 1880 and 1900 attendances steadily improved in all types of Bradford schools, although those controlled by Board schools remained highest throughout the entire period. Evidently, the Church of England schools had inferior attendances when compared to the new Board schools created throughout the period 1880-1900. The gap only narrowed between the two schools on one occasion in 1892 as the result of an extended period of cold weather in the town.

The Triennial Report stated that there was 'much distress in the town caused by the frost'.[30] Such was the distress caused that the 'school attendance officers assisted the Charity Organisation Society'.[31] When visiting the homes of poor attendees they assessed each family's socio-economic situation, the deservingness of those on the cusps of relative-to-absolute poverty, and forwarded that information to local charities, who provided some relief where it was judged necessary.[32] A general rise in attendance rates between 1880 and 1900 cannot, however, simply be explained as being attributed to one single socio-economic factor. What is known is that there were various amendments to education legislation and these numerically increased the numbers of Bradford attendance officers, all of which played a part. Legislative amendments included the raising of minimum standards and minimum attendances before part-time working was allowed, and also raised school leaving ages. It is important thus to compare the Bradford's average attendance rates against the national average.

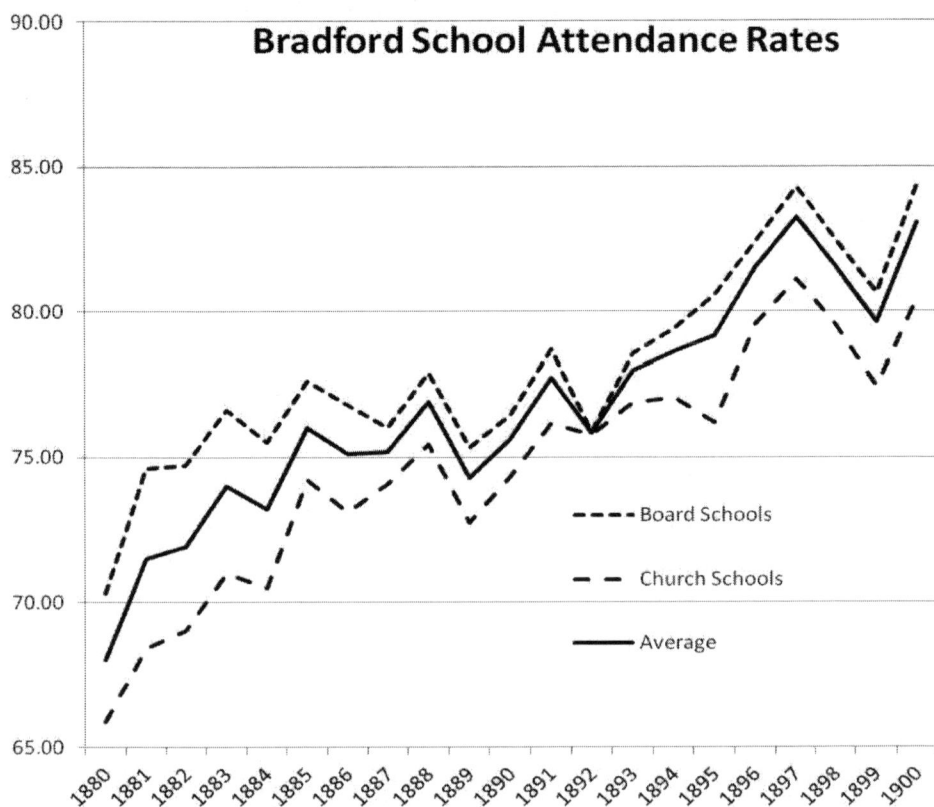

Fig. 2. The percentage attendance of children at Board and Voluntary Schools in Bradford 1880-1900
Source: Bradford Local Archive Library[29]

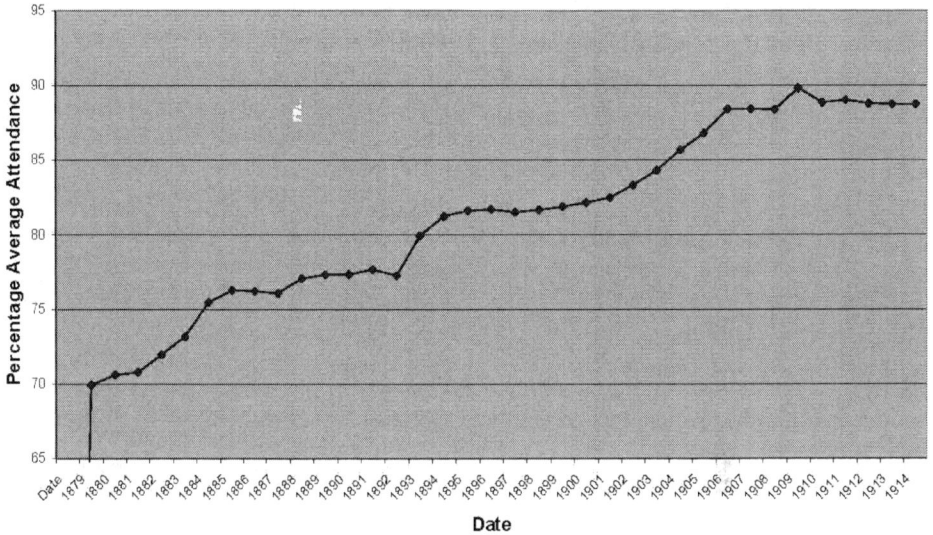

Fig. 3. National Elementary School Attendance in England, 1879-1914
Source: Sheldon, 2007, p.50

Figure 3 shows the national attendance trends using data obtained from the Committee of Council on Educational Reports collated from 1879 to 1899.[33] When this data is compared with Fig. 2, it shows the Bradford average was lower than the national average by a mere one or two percent. Yet, for the local Church of England schools this was closer to four per cent. These findings are consistent with past studies by Sheldon, *et. al.*, that have showed that Bradford's average attendances were always lower than the national average. In order, therefore, to determine whether the schools in this study followed the Bradford trends, a survey of their results will be compared against those in Table I.[34]

TABLE I
Average Attendance Data for the Board and Church Schools at Bradford, *ca* 1885 to 1900

Board School	1885	1888	1891	1894	1897	1900
Accommodation	304	810	810	810	810	810
On Register	342	390	427	442	536	557
Avg Attendance	266	284	325	363	452	486
% Attendance	77.78	72.82	76.11	82.13	84.33	87.25
Church School	**1885**	**1888**	**1891**	**1894**	**1897**	**1900**
Accommodation	209	209	209	209	209	209
On Register	271	268	238	186	225	227
Avg Attendance	211	196	180	136	186	193
% Attendance	77.86	73.13	75.63	73.12	82.67	85.02

Source: BLAL, REP 379, Bradford School Board Triennial Reports, two volumes,
data collected from 1882-1894, and 1895-1903

The data in Table I shows that the Church School attendances were slightly better than the Board School in 1885 and 1888; afterwards, however, the situation was reversed. In contrast with Fig. 2 the Chapel Green Church School only exceeded the overall Church School average attendance in Bradford during 1885 and 1900. Meantime the local Board School being investigated exceeded the Bradford Board School averages in 1885, 1894, 1897 and 1900. There was a similarity in absence levels when the two schools were housed in the same building. However, after the Board school moved to a larger property in 1886 their attendance rates improved. Thereafter after 1888 they were always superior to the Chapel Green Church School. The reason for this now needs further exploration.

The School Environment and Absence Levels

To assess whether the school environment played a part in absence levels it is necessary to explore the physical state of the school buildings. Staff and pupils might not have attended if the structure, maintenance or layout of the premises were sub-standard. There is though one caveat to keep in mind, since the situation is at times difficult to assess as the logbooks only record the most serious fixture and fittings issues. This might have something to do with not wanting to place teachers' employment in jeopardy if the buildings were judged to be so bad they should close. Early in the period under discussion the following logbook entry states that 'in several places the rain drops through the roof [and] the wind has blown a pane of glass in the Infant room making the room very uncomfortable for both scholars and teachers'.[35] A surprise visit by His Majesty's Inspectorate (HMI) reveals the inspector found that the 'school offices are in a dirty and unwholesome state'.[36] One report recorded a 'wretchedly bad building' but noted the Board 'intends to provide better buildings immediately'.[37] There is no further mention of the state of the Church School Building for more than ten years until 1896 when the school was thoroughly cleaned. One effect of poor building maintenance was the low temperatures that staff and pupils of the Church School experienced during the winter months when the outside temperatures plummeted. Temperature is mentioned several times. By way of example, inspection report entries state: 'it has been very cold in school during the week'[38]; 'school very cold, much trouble to get either coal or coke'[39]; and 'the school is terribly cold, the children's fingers have been quite numb'.[40] The Board School Logbook, although uncomplimentary about the building, does not single out specific defects, although it too mentions the cold environment. Both school logbooks revealed that on occasion fires were lit, presumably when it became too cold to function efficiently. Hot days were also problematic with the Church School letting 'a class work in the yard because of the heat'.[41] The Board School likewise recorded that 'the heat has been felt much in school' and that 'several have gone out of school sick, chiefly I believe for this reason'.[42] It should be noted that after the Board School moved to its own premises, these comments stopped. Unlike the premises of the Congregational Church, the new school had larger classrooms and was very spacious in comparison: 'Each room has its own fresh air supply, independent of doors and windows', and 'the rooms are warmed by hot water apparatus'.[43] The Congregational Church building was evidently poorly maintained and subsequently these conditions can be linked to the main causes of repeated absence for both staff and pupils, particularly when cold. The lack of suitable ventilation in the school would have exacerbated the spread of disease, as shown by Hardy in her epidemiological investigation on the behaviour of infection disease in the Victorian city. The use of one fireplace upstairs and downstairs to heat the building could not be described as fit for purpose and the fumes from the burning coal would not have been pleasant or congenial for those with childhood asthma or suffering from pulmonary infections of the lungs – common conditions at the time.

As a consequence of the building layout and rising numbers, overcrowding became an issue in maintaining school rolls. As a consequence, it had an effect on school finances and health. These are illustrated in the annual government reports which stated: 'this school is in an overcrowded state'[44] and 'if the overcrowding continues I shall recommend a serious deduction of the grant'.[45] The classrooms in the Board school were also examined with the HMI stating that there 'must be no more overcrowding in small classrooms' as it was 'unhealthy'.[46] These rooms had a limited capacity – the recommended accommodation level was a maximum of fifteen pupils. It was thus noteworthy that an unannounced visit by the HMI found class sizes of forty-four and thirty-eight. Clearly the Board and Church schools had overcrowding issues and this contrasted with average attendance levels in Bradford per se, which appeared to be under the recommended capacity limits.

Staffing levels within a school were a crucial part of its success. The School Board Triennial report also recognised their role when it stated: 'experience shows that good attendance is in a degree dependent upon the teachers'.[47] The logbooks reveal differences in staffing levels between the two schools, although there is some disparity concerning the detailed level of information available. For example, the Board School Logbook recorded staff names regularly, while Church School Logbook did not. Understaffing at the Church school, however, is recorded elsewhere and this suggests that because of the substandard conditions described above, it had been difficult to recruit staff. Indeed one emblematic example appears to confirm this observation: 'this school [the Anglican one] has been starved of adequate staff'[48]... and 'the staff is not sufficient for the number of children attending'.[49] One HMI reports significantly states that 'the staff continues to be totally inadequate to carry on this school with efficiency'.[50] Poor staffing levels hence resulted in the Church School effectiveness being regularly rated as bad or poor. The Board School also suffered from staffing levels, which were below the recommended levels. This however did not present a major concern as it was not mentioned by the HMI, perhaps because the school was regularly passed by the Government Annual Inspection with ratings of fair, good or excellent. That said, staff sickness levels were a matter of concern to both schools. Illness affected their ability to undertake effective teaching and accordingly impacted on pupils. To examine the extent of staff absence due to illness, an examination of the school logbooks reveals data collated in Table II below.

As Logbooks can be inconsistent in the level and quality of information, the data in Table II has been presented as a conservative estimate. Yet, it does indicate that staff absences occurred at both schools. The main causes of absence identified were the common 'cold' and 'severe cold'. It was not uncommon to see a correlation between staff and pupil absence for the same illness. For example a teacher absence was reported as 'suffering with a severe cold[52] at the same time as 'children are absent with colds and sickness'.[53] The length of teacher absences at both schools tended to be between three to five days, although some absences continued for several weeks, particularly at the Church School which was already plagued with staffing issues. Their staff absence had a major effect on the effective running of the school and put pressure on the remaining staff as they had to cover both the classes of absent staff and their own as well. The Board School by contrast covered their long-term absences with a supply teacher. It is noteworthy then then that one entry at the Church School stated an absent Pupil Teacher has: 'been ordered by the doctor to have one or two weeks entire rest'.[54] It is difficult to know whether a stress-related illness had been the root cause of prolonged sickness episodes, but the possibility cannot be ruled out either. The records suggest that stress within the teaching environment was not a modern phenomenon for the rank and file, whereas head teacher absences at both schools were mostly confined to one to two days.

TABLE II
Staff sickness absences recorded in the School Logbooks, 1889-1890s

	Church School		Board School	
	1880s	1890s	1880s	1890s
Head Teacher	18	5	13	8
Other Staff	20	14	19	32

Source: West Yorkshire Archive Bradford.(51)

Examining pupil absences in-depth, Table III displays the reasons stated for pupil absences in the various school logbooks and provides a more personalised picture of those premises featured in this study.

TABLE III
Pupil Board School attendance rates and reasons stated for absence in the school logbooks, 1880s to 1890s

Board School	1880s	1890s	Church School	1880s	1890s
School Fees	1	0	School Fees	4	0
Truancy	0	0	Truancy	0	0
Leisure Events	19	21	Leisure Events	4	14
Sickness	18	16	Sickness	24	10
Weather	25	11	Weather	3	0
Half Timer	0	0	Half Timer	0	0

Source: West Yorkshire Archive Bradford[55]

Table III suggests there were three main causes of absence at both schools – namely, leisure events, sickness and weather, of which the Horton tide and the Wibsey tide were the main causes of this form of absence. Both schools handled this problem differently. The Board school had major attendance problems on either side of the tide days. Occasionally they viewed these days as a holiday but the main strategy was to let pupils out of school early. Regardless of this, a large number of pupils were absent all day. The head teacher showed his frustration when he recorded 'these tides ought to be thrust out of the borough'.[56] By contrast, the Church School gave pupils a week's holiday for the Horton tide and shortened the summer holiday. For the Wibsey tide, when poor attendance was the norm, they closed the school if attendance was very low or gave half-day holiday in the afternoon. The Anglican approach then to tidal events and their leisure patterns was to be more pragmatic than their non-denominational counterparts. Yet, on closer inspection the reverse seems to have occurred during unseasonal metrological events. Absence due to the weather being unfavourable was more of a problem at the Board School, than the Church School. This though is also as a consequence of different styles of writing in the logbooks, rather than one set of pupils being more impervious to the weather. Weather related attendance patterns were nevertheless a common theme in previous studies and can be credited mainly to 'poor clothing and lack of drying facilities at schools'.[57] As we have seen for staffing levels, sickness amongst pupils was a problem for both schools. Zymotic diseases, such as measles, scarlet fever, and whooping cough, were all regularly recorded in the logbooks. Reducing the spread of disease in schools was important and thus children were not allowed to attend if disease was present in a family. This was seen in the logbook comment of 'many of the children absent through measles being in their family'.[58] Equally the Board School followed the same rule and stated: 'others have measles or have them in the family'.[59] The School Board had to be informed of any serious infectious disease at which point the Medical Officer would become involved. His role was to prevent the spread but also to ascertain when it was safe for the children to return to school. Between 1875 and 1889 the only notifiable disease was cholera but after the Notification of Infectious Diseases Act in 1889 the list of common diseases expanded. The Act, however, did not cover the majority of diseases experienced in Bradford. Effectively this meant pupils and parents made their own decision as to when it was safe to return to school. In practice families may have prolonged their absences. Surprisingly there are no references at either school about absences of part-timers, which both schools had. Historians and the Bradford School Board have regularly credited them as being a significant cause of absence. David Rubenstein felt, for instance, that 'the children of the unskilled, the sweated and casually employed went to school unwillingly and stayed there as little as possible'.[60] The School Board reported that 'in the borough of Bradford there is a large number of children employed half time, more in fact than in any other town in the Kingdom, and it is evident that the average attendances must suffer in consequence'.[61] Absences because of gender issues are also surprisingly absent in both logbooks. We need therefore to examine in more detail the HMI reports that complement the information on absences contained in the School Logbooks.

Inspection Reports

The HMI annual school inspection reports are recorded in the actual logbooks and can be very revealing. They contain information relating to the state of the school standards, grants earned and the efficiency of the school at the time of the inspection. The Board School report showed that between 1880 and 1886 comments appeared regularly regarding standards and the building condition. Initially standards were poor and resulted in comments, such as 'great improvement will be expected next year as a condition of an unreduced grant'.[62] The condition of the building was always mentioned in a negative light. Grants were obtained even if reduced and in later years merit grants were sometimes gained. After the school moved to its own building in 1886, substandard building issues stopped being mentioned and standards consistently improved. The logbook records 'the results of the years examinations shows a considerable improvement'; they merited higher grants, which reflected being rewarded for achievement.[63] The Church School, however, had the opposite remarks, with reviews continuing to illustrate consistent poor attainment, grant reductions and negative comments about staffing levels and a substandard building. In 1881 the grant was reduced by 'one tenth for defective instruction in the mixed classes'[64] and the infants received no payment. The HMI occasionally issued threats to try to remedy the situation. An emblematic example read: 'if overcrowding occurs I shall consider a serious reduction in grant next year'.[65] This type of threat was not uncommon for the Church School and, with funding regularly reduced, it is not surprising that it suffered from staffing and building maintenance issues. It is possible to ascertain the financial state of the school from an HMI report that records it should 'be observed that this school is unsupported by either private subscription or aid from societies or collections in church'.[66] Three years later they found 'that the school account is overdrawn by £179'.[67] Despite this shortfall, around this time the school standards began to improve and they started to attain an occasional merit and higher grants for some subjects. The HMI were not ruthless and sought to provide financial help for the school with an aid grant where it was feasible to do so. Consequently, in 1898 the Church School was given a sum of £45 to improve recruitment and to pay for additional staff, apparatus, and furniture. This was followed in the next two years with grants of £49 and £47 to pay for staff and towards the school overdraft.

Without regular fiscal support from the Church of England or by private subscription, educational grants were the only means of viable financial support for the Anglican educational establishment. Although supported by the ratepayers, the Board School still needed external grants to survive. Obtaining Merit and Higher awards was imperative for securing educational grants over the long term. The Triennial report recognised this link between finances and attendance when it stated: 'improvement in attendances effect finance. An increase of 5% would raise the income of the Board from Government Grants about £1500 per year' and 'denomination schools about £750 per year'.[68] The political economy of schools was hence always dependent on improving results and making sure that absences were reduced. Both of these factors were not easy to control when the ethos and physical infrastructure of the School itself was shaped by complexities of broader and changing socio-economic, weather, and epidemiological conditions in late-Victorian Bradford.

Conclusion

This study set out to evaluate if a selection of local schools in Bradford contributed to poor pupil attendance and what that might reveal about conventional views concerning the reasons for poor attendance at late-Victorian schools across the North of England. The new research has provided data that suggests the standard view in the historiography, which stresses sociological explanations for low attendance levels in Bradford and elsewhere, is misleading. Hurt, Ellis, *et. al.*, have all concluded that it was the parent or the child who were at fault for poor attendance at school. Yet, poor attendance rates in Bradford, which have traditionally been determined from official statistics by local historians, are confusing and deceptive. Local data, when compared nationally, appears to show that attendances were poor on location and remained so for the duration of the late-Victorian era. This opinion has then become regarded as an official fact in histories of education. Few historians have sought to question that these trends take into account the fact that officialdom seldom admitted their prejudices or failings in their working practices. Too often, sweeping assessments were formed from statistical data that was collated by averaging trends rather than examining individual schools' logbooks. The findings of this article demonstrate that there were blurred distinctions in which parents could have reasonably claimed that poor attendance was not the fault of them or their child.

It is a noteworthy new finding in this article that the environment in which the children were taught was very important to learning and its significance was eventually recognised by the non-denominational School Board. Bradford Schools and their masters were *in loco parentis* over their pupils but the evidence shows that they often faltered in this role when it came to maintaining and improving the infrastructure of the physical environment to improve teaching standards – classes were often freezing in winter, too hot in summer, or heated by noxious coal fires that exacerbated common conditions like childhood asthma. An additional problem was overcrowding and broken windows, frequently seen in the Church building plan, and recognised by school inspectors who recorded that it was unhealthy but not acted upon by the Church managers. Effectively, the Anglican Church was complicit in allowing the spread of disease. Not only were pupils at risk, but also staff. Evidence shows staff absences can be historically attributed to illness patterns that were exacerbated by these physical conditions. Teaching standards did suffer and were compounded by understaffing patterns, particularly in the Church School. The Board School had similar problems when they shared accommodation with the Church School. Under such conditions illnesses, such as colds and the spread of them, were inevitable and therefore poor building maintenance caused absence to rise in the school itself. Exclusion from school if a pupil was ill or there was illness in the family was a necessity to reduce the spread of illnesses. This was often beyond the control of parents and contributed to absence rates. Data is not available to examine the epidemiology in depth and immunity levels but it is evident that the expansion of public health monitoring did impact on school attendance levels. Whatever the underlying healthcare reason, pupils had enforced absences and it is misleading to conclude from those statistical returns that either the child or the parent caused low attendance levels. An expanded study of a wider cross-section of late-Victorian school premises and their pivotal role in shaping pupil attendance patterns would facilitate a more nuanced appreciation of to what extent national trends in attendance levels reflected local realities. For there is no doubt that school logbooks continue to function in family and community history as a fascinating historical prism into changing educational standards, and their fluid cultural nature, during the late-Victorian era.

Notes

1. Wright, S. Teachers, Family and Community in the Urban Elementary School: Evidence from English School Logbooks, 1880-1918. *History of Education*, 2012, 41:2, pp.155-73, quote at p.157.
2. 33 & 34 Victoria, C. 75, 1870.
3. Best, GFA. The religious difficulties in National Education. *The Cambridge Historical Journal*, 1956, 12:2, p. 162.
4. Ibid.
5. For a recent appraisal of the history of the education voluntary sector at home and abroad, *see*, notably, Aldrich, R. The British and Foreign School Society, Past and Present. *History of Education Researcher*, 2013, 91, pp.5-12.
6. *See* Horn, P. *The Victorian and Edwardian Schoolchild*. Amberley Publishing Ltd, Sussex, 2010.
7. *See* Allsobrook, DI. *Schools for the Shires: the Reform of Middle Class Education in mid-Victorian England*. Manchester University Press, 1986.
8. Hurren, ET. A Radical Historian's Pursuit of Rural History: The Political Career and Contribution of John Charles Cox, c. 1844 to 1919. *Rural History*, 2008, 19, pp.81-103 and Hurren, ET. *Protesting about Pauperism: Poverty, Politics and Poor Relief in late-Victorian England*. Boydell and Brewer, Woodbridge, 2012, p.127.
9. *For example*, Auspos, P. Radicalism, Pressure Groups and Party Politics: from the National Education League to the National Liberal Federation. *Journal of British Studies*, 1980, XX, pp.181-204; Griffiths, PC. Pressure Groups and Parties in late-Victorian England: The National Education League. *Midland History*, 1975, III, pp.19-26; Jay, R. *Joseph Chamberlain: A Political Study*. Oxford University Press, 1981, recount key political debates in educational reform 1870-4.
10. There is an extensive literature on Gladstone's drive for educational reform, *see*, selectively, Vincent, J. *How the Victorians Voted*. Cambridge University Press, 1967; Vincent, J. *The Formation of the Liberal Part, 1857-1868*. Cambridge University Press, 1966; Bentley, M. *Politics without Democracy: Great Britain, 1815-1914*. Blackwells Publishers Ltd, London, 1984.
11. Hurt, J.S. *Elementary Schooling and the Working Classes, 1860-1918*. Paul Kegan and Routledge, London, 1979, p.3.
12. Ibid.
13. Ibid, p. 157.
14. 43 & 44 Vict. C. 23, 1880.
15. 54 & 55 Vict., C. 56, 1891.
16. Bradford Local Archive Library (BLAL) REP 379, Bradford School Board Triennial Report, Volume 1, 1893, 46.
17. Ellis, ACO. Influences on School Attendance in Victorian England. *British Journal of Educational Studies*, 1973, 21, p.317.
18. Hurt, p.31.
19. Thompson, FML. Social Control in Victorian Britain. *The Economic History Review*, 1981, 34, p.191.
20. Humphries, S. *Hooligans or Rebels? An Oral History of Working Class Childhood, 1889-1938*. Brandwell Publishers, Oxford, 1981, p.64.
21. Ibid.
22. Ibid, p.30.
23. Rose, J. Willingly to School: The Working-Class Response to Elementary Education in Britain, 1875-1918. *Journal of British Studies*, 1993, 32, p.114.

24. Roderick, GW and Stephens, MD. *Education and Industry in the Nineteenth Century*. Longman, London, 1978, p.16.

25. Ibid.

26. Best. The religious difficulties in National Education. p. 170.

27. *The Derby Mercury*, 22 October 1862, p.4.

28. Ibid.

29. REP 379, Bradford School Board Triennial Reports, two volumes, data collected from 1882-1894, and 1895-1903

30. BLAL. REP 379, Bradford School Board Triennial Report , Vol 1, 1894 report, p. 43.

31. Ibid.

32. *See* Humphreys, R. *Sin, Organized Charity and Poor Law in Victorian England*. Oxford University Press, Oxford, 1985.

33. The data from 1900-1914, which does not form part of this study, comes from the Board of Education Annual Reports and form part of the graph which I cannot amend.

34. The Triennial Reports only provided data as three yearly averages therefore annual data is unavailable for direct comparison with Fig 1.

35. West Yorkshire Archive Bradford (WYAB), BDP30, Chapel Green Church School Logbook 1868-1903, 1 October 1881.

36. WYAB, BDP30, Chapel Green Church School Logbook, 1868-1903, 19 June 1885.

37. WYAB 65D00, Chapel Green Board School Logbook, 1877-1905: 10 April 1885.

38. WYAB, BDP30, Chapel Green Church School Logbook 1868-1903: 16 December 1881.

39. Ibid., 10 November 1893.

40. Ibid., 1 December 1893.

41. Ibid., 28 June 1886.

42. WYAB, Bradford, Chapel Green Board School Logbook 1877-1905, 4 June 1883.

43. BLAL, REP 379, Bradford School Board Triennial Report, Vol 1:1887 report, p. 16.

44. WYAB, BDP30, Chapel Green Church School Logbook 1868-1903, 16 June 1886.

45. Ibid., 4 June 1890.

46. WYAB, BDP30, Chapel Green Board School Logbook 1877-1905, 11 April 1884.

47. BLAL, REP 379, Bradford School Board Triennial Report, Vol 1, 1894 report, p.39.

48. WYAB, BDP30, Chapel Green Church School Logbook 1868-1903, 19 February 1884.

49. Ibid., June 1885.

50. Ibid., 19 September 1889.

51. WYAB BDP30, Chapel Green Church School Logbook 1868-1903 and 65D00, Chapel Green Board School Logbook 1877-1905.

52. Ibid., 26 November 1889.

53. Ibid.

54. WYAB, BDP30, Chapel Green Church School Logbook 1868-1903, 1 January 1891.

55. WYAB, BDP30, Chapel Green Church School Logbook 1868-1903 and 65D00, Chapel Green Board School Logbook 1877-1905.

56. WYAB, 65D00, Chapel Green Board School Logbook 1877-1905, 5 September 1884.

57. Taylor, S. Bell, Book and Scandal: The struggle for school attendance in a South Cambridge village school 1880-1890. *Journal of the Family and Community Historical Research Society*, 1998, 1, pp.71-84.

58. WYAB, BDP30, Chapel Green Church School Logbook 1868-1903, 12 May 1882.

59. WYAB, 65D00, Chapel Green Board School Logbook 1877-1905, 22 August 1884

60. Rubenstein, D. School Attendance in London 1870-1904. *Occasional papers in economic and social history*, Hull University, 1969, p.112.

61. BLAL. REP 379, Bradford School Board Triennial Report, Vol 1. 1885 report, p. 46.

62. WYAB, 65D00, Chapel Green Board School Logbook 1877-1905, 29 March 1883.

63.   Ibid., 16 March 1888.
64.   WYAB, BDP30, Chapel Green Church School Logbook 1868-1903, 7 March 1881.
65.   WYAB, BDP30, Chapel Green Church School Logbook 1868-1903, 4 June 1890.
66.   Ibid., 29 April 1893.
67.   Ibid., 2 April 1896.
68.   BLAL, REP 379, Bradford School Board Triennial Report Vol 2:1895 report, p. 42.

Bibliography

Primary Sources
Bradford Local Archive Library, REP 379, Bradford School Board Triennial Report , two volumes.
    Volume 1 contains reports, 1882-1894, and Volume 2 contains reports 1895-1903.

Secondary Sources
Aldrich, R. The British and Foreign School Society, Past and Present. *History of Education
    Researcher*, 2013, 91, pp.5-12.
Allsobrook, DI. *Schools for the Shires: the Reform of Middle Class Education in mid-Victorian
    England*. Manchester UP, Manchester, 1986.
Auspos, P. Radicalism, Pressure Groups and Party Politics: from the National Education League
    to the National Liberal Federation. *Journal of British Studies*, 1980, XX, pp.181-204.
Bentley, M. *Politics without Democracy: Great Britain, 1815-1914*. Blackwells Publishers Ltd,
    London, 1984.
Best, GFA. The religious difficulties in National Education. *The Cambridge Historical Journal*,
    1956, 12:2, pp.153-73.
Biagini, EF. *Citizenship and Community: Liberals, Radicals and Collective Identities in the British
    Isles, 1865-1931*. Cambridge UP, Cambridge, 1996.
Biagini, EF. *Liberty, Retrenchment and Reform: Popular Liberalism in the Age of Gladstone, 1860-
    1900*. Cambridge UP, Cambridge, 1992.
Ellis, ACO. Influences on School Attendance in Victorian England. *British Journal of Educational
    Studies*, 1973, 21, III, pp.313-26.
Griffiths, PC. Pressure Groups and Parties in late-Victorian England: The National Education
    League. *Midland History*, 1975, III, pp.19-26.
Hardy, A. *The Epidemic Streets, Infectious Disease and the Rise of Preventative Medicine, 1856-
    1900*. Oxford UP, Oxford, 1993.
Horn, P. *The Victorian and Edwardian Schoolchild*. Amberley Publishing Ltd, Sussex, 2010.
Humphries, S. *Hooligans or Rebels?: An Oral History of Working Class Childhood, 1889-1938*.
    Brandwell Publishers, Oxford, 1981.
Hurren, ET. A Radical Historian's Pursuit of Rural History: The Political Career and Contribution
    of John Charles Cox, c. 1844 to 1919. *Rural History*, 2008, 19, I, pp.81-103.
Hurren, ET. *Protesting about Pauperism: Poverty, Politics and Poor Relief in late-Victorian
    England*. Boydell and Brewer, Woodbridge, Suffolk, 2012.
Hurt, JS. *Elementary Schooling and the Working Classes, 1860-1918*. Paul Kegan and Routledge,
    London, 1979.
Jay, R. *Joseph Chamberlain: A Political Study*. Oxford UP, Oxford, 1981.
Rose, J. Willingly to School: The Working-Class Response to Elementary Education in Britain,
    1875-1918. *Journal of British Studies*, 1993, 32, II, pp.114-38.
Thompson, FML. Social Control in Victorian Britain. *The Economic History Review*, 1981, 34, II,
    pp.189-208.
Vincent, J. *How the Victorians Voted*. Cambridge UP, Cambridge, 1967.

Vincent, J. *The Formation of the Liberal Part, 1857-1868*. Cambridge UP, Cambridge, 1966.

Unpublished Dissertations and Papers
Taylor, AF. 'Birmingham and the movement for national education'. Unpublished PhD., Leicester University, 1960.

Biography

After retiring from work Ray Greenhough attended Bradford University as a mature student and gained a Degree in Local and Regional History in 2011. His main area of interest is Education and he has several projects on this subject that he is pursuing.

A LATE-FLOWERING LOGBOOK
AYLSHAM COMMUNITY NURSERY SCHOOL 1949-1960
Stella Evans

Illustration I. *Bayfield House, Aylsham, Norfolk*
Photo © Gwynneth Starling

School Log Books necessarily tell us about education its methods and teacher training. However, the Log Books for Aylsham Community Nursery School at Norwich Record Office reflect the personal experience of the inhabitants of the town in which it was based. They give a picture of the concerns of families and teachers of the period. Because of the age of the pupils they cover attitudes towards child care and play. They cover family networks and parents' occupations. They place the school within the local community through its connections with local government, medical practitioners and the trades and services of the town. They illustrate ways in which over the last fifty years there have been changes in transport, communications and attitudes to the type of building suitable for education.

Many of the documents relating to the mid twentieth century are still not available to historians. The result is that few studies have been possible in this period. Archivists often apply the criterion of the census-related hundred-year-rule to documents that might name living people, so, it is unusual to find any contemporary description of pre-school childcare and education at this time. Great care has to be taken in writing about people who may still be alive or may still have close relatives living. Conversely, the impact of the developing Welfare State and post war economic recovery on the lives of local communities is of great interest, one that provides a very wide field for study. In an article for the National Archives Dominic Sandbrook says:

> At the beginning of the 1950s, after all, Britain had been threadbare, bombed-out, financially and morally exhausted. Yet within less than ten years, everything had changed; indeed, perhaps more than any other post-war decade, it was the 1950s that transformed Britain's social and cultural landscape. .As the economy began to boom, wages soared and unemployment almost disappeared, everyday life became more comfortable.[1]

This micro-study looks at the logbooks of one small nursery school and considers how they reflect these changes. The books that are the subject of this analysis cover the decade from 1949 to 1960. They form part of a larger collection of twenty boxes, which contain records and photographs of Aylsham Community Nursery School from 1942 to 2001.[2] Some of these later records are, of course, still closed to access by the public, but the logbooks are now open for examination. The vast majority of the entries are written by Miss F Clark, although those for the first year are written by Miss Nicholson, who worked as superintendent while Miss Clark was retraining as a teacher, and towards the end of the decade many of the records are written by Mrs Esme Dain, her successor.

Aylsham

Aylsham is a small market town situated about half way between Norwich, and the north Norfolk coast. It is surrounded by both arable and livestock farm land, and being on a river was ideally placed as a centre for local trade. Evidence of the importance of Aylsham as a trading centre in the seventeenth and eighteenth centuries is confirmed in that there was a turnpike road opened from Norwich in 1794, and about the same time the river Bure was canalised. During this period of growing prosperity the town expanded to the north and west, with several large houses being built between the market place and the mill on the river. Among these was Bayfield House. During the nineteenth century manufacture in Norfolk became concentrated in the three largest towns, causing the county to rely heavily on agriculture. Trade in Aylsham declined. The 'Navigation' deteriorated, its locks finally succumbing to the extensive floods in 1912.[3] By the middle of the twentieth century Aylsham was a stable community with road and rail links to Norwich, the coast and the surrounding villages. Its population in 1951 was 2,526 and by 1961 had increased by only 109 to 2,635.[4] It still has a twice-weekly market and monthly 'farmers' market', the livestock market did not close until the early twenty-first century.

Nursery Education

Attitudes to Nursery education have been a subject of debate ever since the Scotsman Robert Owen first provided facilities for his workers in 1816. There is a continuing assumption that the best place for small children is at home with their mother, but this is not always practicable.[5] Working mothers have always needed to find alternative care for their younger children. Traditionally this was provided by older siblings or child-minders.[6] With the increased concentration of the population in towns in the eighteenth century dame schools became prolific. The innovation that Robert Owen introduced in his school at his Scottish cotton mills was the employment of specialised teachers to supervise the one- to six-year-old children of his employees in an airy schoolroom.[7] In England the early nursery and infant schools were very much the preserve of the churches. The dissenting churches were active earlier in this field than the established church. These nurseries, apart from the emphasis on religious education, had a much more formal and book-based curriculum. One option for mothers of preschool children that became available in the late nineteenth century was to enrol them in the 'babies' class attached to the local Board School.[8] During the twentieth century nursery schools came under the control of the local authorities.[9] Another facility for preschool children was provided by day nurseries. These came under the Ministry of Health, with resulting overlap of provision and consequent confusion. Parents did not necessarily grasp the difference between a day nursery and a nursery school.[10] Teaching methods by the 1950s for preschool children had more in common with the 'child-based' approach of the system, which Robert Owen inaugurated.[11] In her report to the Aylsham nursery school managers in December 1954 Miss Clark states:

> Play with these (propelling) toys is valuable as a preparation for the future life and helps the child to develop his character by the exercise of self-control, self-reliance, patience, and persistence that he brings to his play.[12]

This comment reflects the concern in early years' education that emphasised physical and social development as well as more book-based learning. This latter was not neglected but was encouraged by the use of rhyme and music.[13]

During World War II the 'Wartime Nurseries Scheme' was funded by the government.[14] This scheme was designed to free more women to boost the depleted workforce. Priority was given to areas that received evacuees from the bombing. Aylsham was used by heavily bombed Norwich as a base for evacuated children and families. A logbook entry in September 1954 records a visit from some children from London who had been admitted to the nursery in 1942.[15] Responsibility for these nurseries and nursery schools was initially given to the Ministry of Health, but in 1945 they were handed over to local authorities.[16] The 1918 Education act, section 19, made provision for local education authorities to supply nursery schools for children over two and less than five years of age stating that 'such a school is necessary or desirable for their healthy physical and mental development'.[17] Internal discussions within the Board of Education declared 'the needs of children under five can in general best be met in the small self-contained Nursery Schools'.[18] Nursery school provision was not, however, made compulsory.[19] The 1944 Education Act contained the proviso that nurseries need not be established 'where the authority consider the provision of such schools to be inexpedient, by the provision of nursery classes in other schools'.[20] Under the Ministry of Health it had been decided that the superintendents of nurseries would be trained nursery nurses. The detailed entries in Miss Clark's logbooks reflect this early training, with much of the material in the books concerning the health of the children and treatment of those who appear unwell or have had an accident, as well as reports of the 'periodic' visits of the school doctor and nurse. Between January 1949 and September 1957 there are 185 references to visits from the nurse, and ninety-six from the doctor.[21] With the transfer of the wartime nurseries from the Ministry of Health to the education departments of the local authorities it was considered desirable that the superintendents should retrain under the scheme for Post War Educational Facilities and Training.[22] Miss Clark trained under this emergency scheme. She left for Wall Hall College, in Hertfordshire, in January 1949 and returned in February 1950.[23]

Bayfield House

Aylsham Community Nursery School was set up in 1942, at which time Miss F Clark was appointed as superintendent. It was housed in Bayfield House, a three storey, redbrick building with a large wing at right angles to the rear that follows the line of the Cromer Hill. The HMI report for 1952 states that:

> The slope and surface of this area makes heavy demands on normal wheeled toys and the school needs stronger outdoor equipment.[24]

This is the only summary of an HMI report that appears in the logbooks for this decade.

The children entered the premises into the garden through a pair of gates.[25] The main entrance has only a narrow path between the door and White Hart Street. If the children had entered through this door they would have been waiting on a dangerously busy road. There are several references in the logbooks during 1952-1954 of cars parking on the road at the entrance to the nursery, and when the police provided a 'No Parking' sign they were merely moved to a position where they blocked the side entrance. A second sign was provided. It is an indication of the low level of car ownership at this time that the nursery staff were able to give the police the names of the offenders.[26]

The records of the school include a plan of the site, but this gives no indication of the layout of the building itself. This can only be gained from the logbook entries. There are some twenty rooms mentioned altogether, including the attics, and sheds for storing prams, bicycles and toys. These latter may have been an integral part of the service wing, as mention is made of the passage between the kitchen and adjoining drying shed.[27] With the development of greater awareness of Health and Safety, in 1957 the Sanitary Department objected to the use of the kitchen for laundry purposes. Mr Sparkes, the building inspector, reported that it had been suggested that the laundry outhouse be used for all the laundry work, not just drying, and that a sink was to be installed there for that purpose.[28] Other concerns for safety were also a direct consequence of the design of the building. There are references throughout the whole period of accidents caused by the nature of the 'kitchen stairs' and the uneven nature of the passage in the service wing, and of cleaners and laundry staff slipping on the stone floor of the laundry.[29] There is a reference of a visit from representatives of the County Architect's Department to look at uneven floors and 'the steep step in the babies' bathroom'.[30] Another safety hazard was the use of cinders to repair the garden path. Later certain areas of the garden were treated with shingle, with the consequent twisted ankles and grazed knees.[31]

A look at the function and position of the rooms in the main part of the house shows how the abilities and safety were considerations when allocating them for use by the children. The rooms for use by the 'Babies', children between two and three years of age, were housed on the ground floor. Those for the 'Children' were situated upstairs. The attics were not used, although there is reference to them being cleaned, and the possibility of them being used as a playroom.[32] There is mention of two other 'playrooms', one for the 'babies' on the ground floor, and one for the 'children' upstairs. Other rooms mentioned include a 'rest room', 'dining room' and 'lavatory or bathroom' for the children on each floor. The play rooms at Bayfield House were at the front of the building. This was the southern more cheerful side of the building. The toilets at the back overlooked the garden.[33] The music room was on the ground floor.[34] This meant that it could be used by all the children. Meals were often taken in the garden, and the children, all of whom were under five years of age would carry their own chairs up and down the stairs, a practice that would be condemned by current health and safety concerns.[35] The rooms used for administration are more difficult to locate. There is mention of a 'staff room' and toilets for the staff. Also there was 'Miss Clark's office' and a 'stock room'. This last may have been the 'first floor store room', which in one entry is designate the 'telephone room'.[36] This last entry, which implies that the telephone was not situated in the office, is an indication of the restricted distribution of telephones, as is a reference in 1950 to the local primary school having no designated phone.[37] The use of old, large houses as schools was common practice at the time, and Aylsham Nursery School continued to occupy Bayfield House until it was moved to purpose built premises in 1971,[38] by which time concerns for the safety of small children meant that it was considered preferable that they should be educated in single storey buildings.

Miss Clark

Who was Miss Clark? Any judgement of her character that is revealed by her writing is necessarily subjective. The initial response may be that Miss Clark's entries are repetitive and pedantic. However, these are the very qualities that make her logbooks such valuable social documents. They also reveal her innate kindness and concern for the children in her care. Among almost daily references to the children's health and welfare there are three entries that can be used to illustrate ways in which she was prepared to make an extra effort on their behalf:

> 14th December 1951: (A small girl) was unwell with a temperature 99o she slept in Miss Clark's office.

> 12th July 1955: Miss Clark went to Norwich on the 1.15 omnibus to visit Miss Armes at Horns Lane Infant School. (Another small girl) aged 5yrs 5mnths, [had] attended Aylsham

Nursery School …. Recently she has been disturbed …… [She] was pleased to see Miss Clark and treasured the Nursery Rhyme Book which Miss Clark had taken for her.

20th December 1956: A small boy complained of earache in his right ear. A hot water bottle was filled and [He] ate his dinner on Miss Clark's lap with his head resting on a hot bottle covered with a blanket.

These entries probably suggest her early training as a nurse, as do their reflection of her strong sense of duty. She expected the same dedication from her staff.

These expectations were the source of the only major dispute with an assistant teacher, a young woman who had been engaged while Miss Clark was away on her year's training. The first indication of Miss Clark's unhappiness with this member of staff appears eighteen months after her return, when she writes that certain children have been having tantrums and been unhappy through mismanagement and lack of understanding of the care of under-fives. Then in September, Miss Clark asked her to submit songs, games and stories before presenting them to the children and not to take any organised games upon arrival of children. Miss Clark had overheard Miss Morgan and a group of children singing in ring games made up by her 'I thought I saw a puddy cat' and 'Three blind mice' who were having their tails 'cut off'.[39] At this time it was Miss Morgan's duty to watch the side gate while the children were arriving and to supervise the free play in that part of the garden. This incident could be cited as an example of Miss Clark's strict hospital training, and her unfamiliarity with certain popular songs and nursery rhymes. There are several entries where Miss Morgan was absent for minor illness or with no explanation.[40] There is another entry which reports her as 'unable to travel' after a visit to her sister in Liverpool.[41] These reports may show Miss Morgan's own unhappiness with her employment at this time. Things came to a head at the end of that year when Miss Morgan refused to clean the upstairs bathroom when the member of staff who usually did this was absent.[42] Miss Clark's medical training made her very aware of health and hygiene issues. After three refusals Miss Morgan had an interview with the County Schools' Inspector, and in January she transferred to Kings Lynn.[43]

Another idiosyncrasy of Miss Clark's was her use of old-fashioned terms. Not only did she use 'omnibus' on several occasions, but throughout the documents she never abbreviated the words 'bicycle', 'linoleum', 'refrigerator', or 'telephone'. If she wrote that someone had 'called' she meant that they had visited in person. She used the term 'lavatory basin' for a washbasin installed in the kitchen, and it was not until 1959 that she used the term 'adhesive plaster' instead of 'strapping'. It is not possible to find any details of Miss Clark's life before she came to the Nursery School. Because of the common nature of her surname it is not even possible to discover her place of birth. Local tradition says that her name was Freda, and she did initial any alterations to the text as 'FRC'. Even in the logbook there are very few references to her personal life. There are mentions of her father and sister, both of whom were still alive in 1959. There are also a few references of Mrs Clark 'from Norwich'. The only clues to her having lived anywhere but Aylsham are references to her visiting her dentist in Great Yarmouth.[44] Although I have no intention that this should be an oral history, one former pupil's only memory of her was of 'her glasses and her smile'.

Staff

One indication to Miss Clark's character can be gleaned from the loyalty and ambition that she inspired in her staff. All the teaching staff regularly attended lectures in Norwich arranged by the Nursery School Association. The HMI report for 1952 is very complementary in its appreciation of the standard of work at Aylsham Nursery:

> The present staff consists of a qualified superintendent, one qualified teacher, a Nursery nurse and a Nursery assistant, all of whom share in the care of the children during the day.
> The domestic staff consists of a cook and a canteen helper, a part-time cleaner and a part-time laundress.
> The enthusiasm of the staff is well expressed by their constant care and provision of equipment.
> The entire staff impresses by their genuine concern for the children".[45]

Miss Clark tended to employ local teenagers and young married women, some of whom had children attending the Nursery. One young girl, who probably came straight from school at fourteen in 1949 was still with her in 1960. While her companion, also first mentioned in 1949, left to train as a State Enrolled Nurse.[46] Mrs Dain, who succeeded her in 1960, first started work at the Nursery in 1950. She deputised in Miss Clark's absence, both in charge of the Nursery and at governor's meetings. She was always willing to go to assist at other schools when pupil numbers were low due to illness. She is variously referred to as 'Nursery teacher', 'assistant teacher' and as 'qualified assistant teacher'.[47] Before taking over in 1960 she attended a six-day course at the University of St Andrews. Likewise, her colleague, Mrs Hudson, returned to further education in 1960, first taking her GCE, and then being accepted at Norwich Training College. The only difference of opinion with these members of staff regarded the wearing of overalls by the teaching staff. In the autumn of 1955 the Norfolk Education Committee asked that all staff should wear overalls, for 'hygienic and aesthetic reasons'.[48] Miss Clark complied, but six months later there is an entry that she reminded the other staff of this.[49] It was another year before there is an entry in the logbook recording that Miss Clark had requested the staff to wear overalls several times.[50] One member said that she could not afford them, while another said 'She would wear them if they were provided by the NEC and the laundry bill was paid by them'. Miss Clark remarks that both assistants were needlewomen and made their own clothes.[51] This was common practice at the time, 'clothes were often homemade, either sewn or knitted'.[52] It was the end of March 1957 before there is a pencilled note in the logbook to the effect that one member of staff wore an overall for the first time, 'Is she reformed?'.[53] There is no further reference so perhaps the matter was resolved amicably.

The first mention of overalls being worn by the children was in 1950.[54] There are several mentions in the books of local dressmakers collecting and delivering material and finished garments, until 'A decision was made by the managers to discontinue the provision of overalls for the children' and the parents would be expected to provide their own.[55] Another concern over hygiene involved the practice of catering staff smoking in the kitchen. At a time when most people smoked there was still recognition of the dangers of contamination to food if the practice occurred in catering areas. As early as 1952 the 'canteen inspector' had voiced concerns about this practice.[56] On four occasions Miss Clark asked the cook to refrain.[57]

It is perhaps easier to get a view of the private lives of the other members of the staff, if only from the reasons given for their absence. Family illness, weddings and funerals, older children's sports days and prize giving, Sunday School outings, and various courses are all recorded. One of the canteen helpers married the cook's son in 1951. She initially left her post but she returned as a temporary cleaner in 1956 and later as a relief cleaner.[58] Mrs Hudson was very active in the Red Cross.[59] Mrs King who did not join the staff until 1958 was the wife of the local Salvation Army band master.[60] Many of these activities reflect the close integration of life in a small town. Before joining the staff Mrs King had been a voluntary helper; an indication of a change of policy by the local authority as previously relief staff had been paid, and the only volunteers had been students who wished to get practical experience.[61]

There seems to have been more difficulty in recruiting cleaning staff, especially towards the end of the decade when the economic recovery caused higher expectations of working conditions. In December 1957 Miss Clark reported to the governors that the current cleaner had said that she could not get through the work in the time allotted. She had raised this with the domestic supervisor, who had pointed out that hours worked were arranged by the Ministry of Education, and the matter had now been referred to the Deputy Education Officer.[62] This difficulty was still not resolved by July 1960. The situation may have been aggravated by Miss Clark's hospital training with its emphasis on hygiene. There are references to her having 'scrubbed' certain items herself when cleaning did not meet her high standards.[63] Mrs Bloom, the cook, was with her for the whole period. She was another local woman who was already working as canteen assistant.[64] Three previous cooks had already resigned; one at least declaring that 'she resented not being allowed to make out the menus for dinner and tea'.[65]

During World War II, technological innovations in the home, and universal education, had changed social attitudes in ways that affected family life, and which were relevant to the care of preschool children. In the 1950s the mother was regarded as the primary carer. In an online article for History Magazine Ellen Castellow says:

> Very few women worked after getting married; they stayed at home to raise the children and keep house. The man was considered the head of the household in all things; …. It was still unusual for women to go to university, especially working class women. Most left school and went straight into work until they married.[66]

The teaching staff of Aylsham Nursery School differed from this general picture in that they all embraced 'the broader expansion of education'.[67]

Parents' Occupations

One of the ways in which the logbooks indicate social changes in this decade is in the references to the children's parents' roles and occupations. This was a time when the extended family was generally replaced by the nuclear family with fewer siblings, which meant that opportunities for social contact for young children became less.[68] In Aylsham support from the community and the extended family continued, but, as the decade progressed, it became more usual to contact the father of the child rather than rely on sibling support. There are several references made of fathers being responsible for bringing children to school. The continuation of the idea of the mother as the primary carer is reflected in the occasions when two of the references later in the record refer to fathers being given letters of 'explanation' for the child's mother.[69] The extension of telephone ownership also contributed to the ease in contacting parents, especially if they were at work. There are references made to fathers being contacted in order to get a message to their wives. This is another indication of the low instance of ownership of phones in the home. Over the whole period, reference is made to members of staff going to fetch the mother of a sick child.[70] Throughout the logbooks there is evidence of sisters taking responsibility for younger siblings. Although in the later records this seems to refer to married sisters or the mother's sister rather than teenagers.[71] There are just two instances of a grandfather being the person who was turned to for extra child care[72] and again only three mentions of children being taken home by older boys.[73] In two of these cases the boy was an elder brother, but the last refers to an uncle, aged fourteen, taking a child to be cared for by her grandmother. Grandmothers occur most often as secondary carers, not only caring for the child when mother was at work, but also being available to fetch a sick child from school, taking them to the doctor's surgery, or attending a school medical examination. There is also one entry in the school Admissions Book where a child did not eventually come to the nursery because her grandmother was against it. The reason for this objection is not given, but it was presumably on the grounds that she was too young for school.[74]

There were, of course, other reasons than having both parents at work for a child attending the Nursery School. Some were by personal choice, especially in the case of professional parents. There is one case that obviously was recommended by the local authority, where the mother is noted as being unmarried and out of work,[75] while others were under the supervision of the Dr Barnardo's charity.[76] There are also several references to foster children, although this again may have been the personal choice of the foster parents.[77]

It is in an analysis of the occupations of the children's parents that best illustrates the structure of life in a small market town in the mid twentieth century. There are two sources from which it is possible to identify the occupations of the various parents. From August 1955 until her retirement in August 1960, Miss Clark made a note in the Admissions Book of the occupation of the fathers and some of the mothers of new children. Other references to the parents' occupations occur in the text of the logbooks. At this time most women conformed to the assumption that women looked after the home.[78] The occupations of married women identified in these documents are mainly in the service sector. They worked in education, in retail, or at the local hospital. The working class women worked in agriculture or as cleaners. Occupations in education include members of staff at the Nursery itself and other local schools. They include a mother who was a domestic science teacher,[79] and the headmaster of the Red House Farm School at Buxton.[80]

Those mothers who worked in retail were listed as shop assistants. Reinforcing the accepted gender roles of the period was the fact that most fathers held more senior positions. They were variously named as a grocer, a shop manager, and as the owner of a 'general stores' in a neighbouring village. The fish-shop owner and licensee of the public house in the same village also sent their children to Aylsham Nursery. Two of the publicans in Aylsham itself had children who attended. Other occupations in the service sector included the local policemen, three of whom were stationed in Aylsham.[81] The fathers of two children were postmen, and the text of the logbooks show that at least one mother worked as a post office clerk.[82] Both the parents of one child were employed by Messrs Purdy, the local firm of solicitors, and a junior member of staff left the Nursery to work there.[83]

Two other large employers were at the RAF airfield at Coltishall, and the local hospital, St Michaels. Four children had 'flying officer' listed as their father, and another named as a 'clerk of works' for the Air Ministry. The father of another child was also probably a member of the Armed Forces; Miss Clark wrote that when a mother 'called to say her husband was against sending David to the Nursery. ... Miss Clark explained the benefits to be gained from sending an only child who was learning two languages, German and English, from his parents'. The mother decided to send him.[84]

The members of the hospital staff who sent children to the Nursery cover the full range of the social classes. They included: two female domestic staff; and three fathers, one was a male nurse, another was a hospital porter and the third was a hospital administrator. The dental mechanic was probably employed at the Ian Sears Clinic on the Norwich Road. As could be expected in an agricultural area, a large number of parents worked on the land. The Nursery had to have at least one member of staff on duty by 8am or 8.30am so that mothers could be in time for the lorries taking them to work in the fields.[85] Other references to women's jobs in agriculture include Roofes greenhouses, the 'pea hut' at Haveringland, and Westwick 'canning factory'.[86] Also closely connected to the agricultural industry would have been Barclay and Pallette's mill, which still operated on the river.[87] The Admissions Book names various fathers as 'farm labourer', 'agricultural labourer', 'cowman', and 'farm manager'.

Shops and Services

Several local shops are mentioned in the text of the logbooks. These form a good way of estimating the trades and professions that would have served a town the size of Aylsham in the middle decade of the twentieth century. They were all owned or managed by local people. As was common at the time 'Most shops were family businesses and traditional in character'.[88] Some of those mentioned continue to trade in the early twenty-first century.

The only shop that was part of a national chain was the International Stores, one of three grocers mentioned. These would, by the end of the period, have become 'self-service' shops, the predecessor of the modern supermarket. Other food shops mentioned were two greengrocers and a fishmongers. Milk was delivered daily, including the Nursery's quota of one-third-pint bottles that were provided by law for the children. Other shops included were the chemist, a newsagent, and furniture, shoe and haberdashery stores. There is no reference to a clothes shop or tailors. Although the haberdasher may have stocked some clothes it would primarily supplied materials for making homemade clothes.[89] The jeweller and clock-smith was opposite the school. There was an optician, two electrical shops and an ironmonger. This proliferation of local shops and services is indicative of a time when shopping was an occupation done on foot.[90] A time before international corporations concentrated their activities in shopping malls and on internet websites. One craftsman who must have been unusual even in 1956 was Mr Johnson the tinsmith, who mended an eight-pint saucepan and a pudding sleeve for the Nursery.[91] Most of the shops in Aylsham, then as now, were concentrated around the market square and in Red Lion Street, which led to the school.

Other services used by the school include four local builders and two carpenters. Mr Hudson, the plumber, was the husband of one of the teaching staff. This is another example of the close-knit nature of a 1950s small-town community. The gas fitter, electrician and piano tuner called regularly, and there were deliveries of coal, coke and kindling throughout the winter months. The use of Mr Cooper's taxi service increased towards the end of the decade. There is only one mention of this being used during the early years, when a member of staff injured her ankle.[92] Later there are several occasions when the taxi was used to transport sick children home. Before 1956 it was more usual for them to be pushed home in 'Miss Clark's pram',[93] or even on the back of Mrs Hudson's bicycle.[94]

Transport

For trips further afield it was most common to use the bus service. The bus between Norwich and Aylsham seems to have run every hour. On at least two occasions sick children were taken home by bus.[95] Also Miss Clark came in the morning by bus.[96] Another reference indicates that she lived on the Cawston Road, which meant that she may have lived some way outside the town.[97] Children from the neighbouring villages would also have come by bus. The logbooks reflect an increase in private car ownership during this time. Early in the records the only reference to cars was for professional or business use.[98] After 1953 there are more references to parents ferrying their children by car, and by 1959 one member of staff was receiving driving lessons.[99] Another indication of the increase in the ease of travel is that in 1949 reference is made to the difficulty of getting a child to the Nursery from Cawston, a village four miles from Aylsham,[100] while by the end of the period children were attending from Coltishall, which was twice the distance.[101] Perhaps the most notable reference to parental car ownership was the occasion when Miss Clark slept with a small child at the Nursery because the father's car was stuck in a snow drift. This was the occasion when the local newspaper reported that Aylsham was without electricity for twenty-nine hours.[102] The majority of references to the weather and its effect on attendance at the Nursery were extreme events, gales, snow, rain, 'intensive heat'.[103] However, there is one pleasant note that shows how warm the Spring was in 1957, because on 11 March 'The children had their meals and played in the garden for the first time this year'.[104]

Costs and Charges

The decade from1949 to 1960 was one when Britain grew out of wartime austerity to a much more comfortable standard of living. This was reflected by an accompanying inflation in costs and incomes. The Bank of England estimates the rate of inflation up in the 1950s as 4.3%. This is nearly twice as much as in the preceding decade.[105] There are very few mentions of staff income in the logbooks. It would appear that the two qualified teachers were paid monthly, but that the unqualified nursery assistants and the domestic staff were paid weekly. The nursery assistants were on a sliding salary scale. One who started in 1951 on a salary of £198-£250 was earning £329.10s six years later, while her colleague, who was still a teenager, earned £60 a year less.[106] There is considerably more information as to the income of the domestic staff. They were employed at an hourly rate, and their hours varied with individual contracts. The rates of pay were calculated by the strange sum of one eighth of a penny. For instance, in 1957 all domestic staff received an increase of one eighth of a penny per hour.[107] In order to get a full week's work of forty hours one employee combined the tasks of canteen assistant and cleaner. She was paid one and a quarter pennies per hour more for cleaning, but could earn more as a catering assistant as the hours were longer.[108] All hours, salaries and wages were regulated by the Ministry of Education.[109] It is in these calculations that the evidence of inflation is most visible. Between 1952 and 1958 a cleaner's hourly rate had risen from 1s 11^1/$_2$d to 2s 11^3/$_4$d. This was well above the calculated rate of inflation.

Another indication of rising costs can be found in the references made to the food bought for the school meals. Until June 1951 the children were provided with a main midday meal, and two 'subsidiary' meals, breakfast and tea. After this the County Auditors Department told Miss Clark to enter only one 'subsidiary' meal on her forms.[110] The cost of this 'subsidiary' meal in 1951 was 3^1/$_2$d a day.[111] In July this was increased by ^{1}/$_2$d a day.[112] The price paid by the parents reduced for younger siblings when more than one child attended the Nursery.[113] By 1956 parents were being charged 10d a day for dinner and 5d for 'subsidiary' meals. Again this was above the estimated rate of inflation. Norfolk County Council was making a profit of ^{1}/$_4$d on each main meal.[114] The meals supervisor was concerned that the midday meal should not cost more than 9^3/$_4$d per day: 4^1/$_2$d meat, 3d gravies and milk, ^{3}/$_4$d fruit and jams, ^{3}/$_4$d potatoes, and ^{3}/$_4$d greens.[115] A meat allowance of 11oz per week was made for each child.[116]

There is another indicator of social change in the matter of diet and food preservation occurs in the entries for November 1954. At this time a refrigerator was provided for the Nursery and frozen fish is mentioned for the first time.[117] This coincided with the end of food rationing.[118] Until then the school needed meat permits from the Ministry of Food.[119] This Ministry also supplied nutritional supplements for the children. Throughout the period both orange juice and cod liver oil were regarded as necessary in order to provide additional vitamin C and D to the children's diet. On one occasion 'the orange juice supplied by the Ministry Of Food was exhausted', and 'Miss Clark brought Robinson's Prepared Barley Water and 'Delrosa' Rose Hip Syrup for the children to have after rest-time'.[120] For children who were considered to need extra nutrition and vitimins by the school medical service, cod liver oil and malt were provided in the form of either

Virol or Maltolive and Iron.[121] In September 1956 the milk provided for each child was reduced from ²/₃ pint to ¹/₃, and subsequently cocoa was made with dried milk.[122]

Medical Matters

Minor ailments are a constant theme throughout the logbooks, usually first aid delivered by the staff was sufficient treatment, but occasionally children needed referral to a doctor or the Jenny Lind Hospital in Norwich.[123] Miss Clark often 'advised' a mother to take a child to visit a doctor. From the tone of the entries it appears that Miss Clark's 'advice' and 'explanation' to the mothers may have been rather emphatic, and she was not above using the same tone of voice to the local doctors. When in 1950 one doctor queried the necessity of a child being referred to him before returning to the Nursery, Miss Clark records that her 'suggestion' had been 'a precautionary measure, the child had a cough and there was whooping cough in the district'.[124]

The school doctor visited the school regularly. An entry for April 1949 refers to 'the annual medical examination',[125] but after 1954 a note of medical examination by the doctor is made every six months. The mothers were invited to these inspections where 'The children were weighed, measured, and given a good overhaul by the doctor'.[126] The school nurse visited more often. Despite the tradition among children of that generation who called her 'the nit nurse' it appears that she was more concerned about dandruff or cradle cap in these preschool children.[127] Olive oil was used to treat these conditions as well as many minor problems, such as removing grit from a child's eye.[128] Several entries in the logbooks reveal that GPs at that time performed minor surgery that now would be referred to the hospital service. There are several entries of swellings being lanced and cysts removed at the surgery. Other instances occur of stitches being inserted by the doctor, sometimes at the patient's home.[129] The mid twentieth century was a time when, before the development of some vaccines, childhood infectious diseases were rife. The Aylsham Nursery logbooks for every year record instances of mumps, measles, chicken-pox and whooping cough. One entry is for immunisation against the latter.[130] Immunisation is mentioned throughout the logbooks, but in the earlier entries do not state the disease. In 1955 polio vaccine became available to children between two and nine years of age.[131] After this the entries specify either diphtheria,[132] or polio.[133] The vast majority of these were administered at the nursery. Another theme that can be traced in the books is the development of the NHS and advances in medical treatment. References to attendance at hospitals in Norwich become more frequent in the latter years. Medically, penicillin had become to be regarded as a cure all. The development of its use from being freely available to prescription drug can be traced in the logbooks. There are several references to both penicillin ointment and chewing gum. The ointment was left by the school nurse to be used for burns, sores and 'rough spots'. The gum could be bought from the chemists, and was used for sore throats. As late as 1956 the nurse applied a 'penicillin plaster' to a septic sore.[134] Penicillin eye drops, tablets and later emulsion were all prescribed by the doctor. However, by 1955 the need to keep a certain level in the blood stream seems have been appreciated as the doctor advised that tablets be taken in jam three times a day.[135]

Toys

As is to be expected in an establishment dedicated to the education of children under five, toys were considered to be teaching aids in themselves. They were regarded as an essential aid to the children's early development. Miss Clark reported to the managers in 1954 'Play is serious to the child and affords education of the most complete kind for body, mind, character and personality'.[136] The toys mentioned show that playthings for young children have not changed very much over the years. Throughout the period the Nursery relied on the generosity of people to donate toys to the school. At a 1954 meeting it was reported that 'We have made an appeal for gifts of second hand pushing toys mainly for the 2 year olds, and have to date received books and toys from parents'.[137] Miss Clark herself was not above donating items to the Nursery. At various times she provided a see-saw,[138] two dolls' prams, and a cuckoo clock. In 1952 Mr Dain came to see the doll's house upstairs with a view to making its furniture.[139] The items donated were not always designed as playthings. In 1952 Mr Smith of Aylsham Motor Co. gave some motor tyres, an inner tube and a steering wheel.[140] While a photo in the North Norfolk News in 1950 was accompanied by the caption 'Given two or three old boxes and a plank, it didn't take these children long to make an imaginary ship'.[141] Later it is reported that workmen were sent to set up a rubber tyre on a chain as a swing.[142]

The garden was well equipped by the education department with a sand pit, paddling pool, climbing frame, swings, footballs, wheelbarrows, scooters, tricycles, and skipping ropes, all of which would help the physical development and coordination of their users. The swings came in various sizes. Was the pulley swing using a mechanised pulley and spindle a zip wire, or was it designed for the staff to swing the very small children in an enclosed seat?[143] Indoors there was a greater variety of toys than would have been found in the normal family home at the time. Among those mentioned are soft toys, dolls, and plastic animals. There were motor cars, lorries and trucks. The Wendy house contained play laundry equipment. There are surprisingly few reports in the logbooks of accidents with toys. One small boy fell off a wooden horse,[144] two instances of children being hit when another child threw a toy,[145] and three of fingers caught in the Wendy house door.[146] On several occasions the wooden toys were sent to Mr Jonas, the carpenter, to be repaired,[147] and on one occasion a large doll was taken to the dolls' hospital in Norwich to be mended.[148] This practice reflects a culture where it was more common to repair possessions rather than discard them if damaged.

The Aylsham Nursery School was not a play group, and it was part of the staff's duties to 'tell stories and take organised play (physical exercises and singing games) indoors and outdoors'.[149] The music room contained a piano, which was tuned every six months. At the Christmas party 'the children sang and acted rhymes and sang carols for the parents and guests'.[150] Then one of the last entries records that a gramophone was delivered from Education stores.[151] Among the teaching aids mentioned are nursery rhyme picture cards, and craft equipment.[152] Mrs Dain seems to have been particularly interested in craft work. In 1955 she 'decided to introduce a new occupation by using Reve's 'Temperablocks' for the children's painting'.[153] Later that year Mr Lord, the Arts and Craft Organiser for Norfolk, showed her how to make angels and snowmen from match boxes.[154] After she became supervisor she continued to expand the children's activities outside the school. The children were taken to see Harvest Festival decorations in the church,[155] and on another occasion twelve children were taken to Pages store to see 'Mr Pastry'.[156]

Conclusion

The logbooks for Aylsham Community Nursery School at Norwich Record Office reflect the personal experience of the inhabitants of the town during the 1950s. They give a picture of the concerns of families and teachers of the period. Because of the age of the pupils they cover attitudes towards childcare and play. They cover family networks and parents' occupations. They place the school within the local community through its connections with local government, medical practitioners and the trades and services of the town. The debate about preschool education has moved on since the period under discussion in this essay, in the twenty-first century there is much more agreement about the fact that early years' training is essential for the later development of children.

Notes

1. http://www.nationalarchives.gov.uk/education/resources/fifties-britain/ ?show=all#introduction.
2. Norfolk Record Office (NRO), C/ED 138.
3. http://www.norfolkmills.co.uk/aylsham-navigation.html.
4. Census Reports, Norfolk.
5. The National Archives (TNA), ED 102 6, in Palmer 2011, 140.
6. Whitbread, N. *The Evolution of the Nursery-Infant School.* Routledge Kegan and Paul, 1972, pp.6-7.
7. www.Historic UK.com: /Robert-Owen-museum; http://robert-owen-museum.org.uk/New_Lanark_Schools.
8. NRO, Y/ED S/35, 11/4/1902.
9. Education Act 1918, clause 19, http://www.legislation.gov.uk/ukpga/Geo5/8-9/39/contents.
10. Palmer, A. Nursery Schools for the few or the many? Childhood, education and the State in mid-twentieth century England. *Paedagogica Historica*, 2011, Vol. 1-2, 142, p.151.
11. Whitbread, 1972, pp.8-10.
12. NRO C/ED 138: Managers' Minutes, 6/12/54.
13. NRO CED 138: Log Book 2/12/52, 12/4/53, 12/7/55.
14. Whitbread 1972, p.101, Palmer 2011, p.152.
15. NRO C/ED 138 Log Book: 9/09/54.
16. *Hansard*, House of Commons Debate, 9/3/45 HC Deb 09 March 1945 vol. 408 cc2425-50, http://hansard.millbanksystems.com/commons/1945/mar/09/war-time-nurseries accessed 2014.
17. Education Act 1918, C. 39 19.
18. Palmer, 2011, p.140.
19. Whitbread 1972, p.103-4.
20. Education Act 1944, Clause 8.2 http://www.legislation.gov.uk/ukpga/Geo6/7-8/31/contents.
21. NRO C/ED 138; Log Book: 29/1/49. 30/9/57.
22. *Hansard* 22/7/43, http://hansard.millbanksystems.com/written_answers/1943/jul/22/war-time-nursery-wardens-educational; *Hansard* 21/11/46.

23. NRO C/ED 138; Logbook: 14/1/49, 27/2/50.
24. NRO C/ED 138, Logbook April 1952.
25. NRO C/ED 138; Logbook 23/7/52.
26. NRO C/ED 138, Logbook 10/06/52.
27. NRO C/ED 138, Logbook: 26/4/52, 25/11/54, 29/11/54, 7/12/54.
28. NRO C/ED 138, Logbook 28/03/57, 11/04/57.
29. NRO C/ED 138, Logbook 2/5/50, 29/5/52, 27/11/56.
30. NRO C/ED 138, Logbook 14/5/52.
31. NRO C/ED 138, Logbook 3-4/5/56, Jan-March 57, 2/5/58.
32. NRO C/ED 138, Logbook: 22/10/50, 21/11/50, 1/12/59 4/7/57.
33. NRO C/ED 138, Logbook 31/12/56.
34. NRO C/ED 138, Logbook 19/2/53.
35. NRO C/ED 138, Logbook 24/4/59.
36. NRO C/ED 138, Logbook 31/3/60.
37. NRO C/ED 138, Logbook 31/01/50.
38. NRO catalogue.
39. NRO C/ED 138, Logbook 25/08/51.
40. NRO C/ED 138, Logbook 29/8/50, 19/10/50, 6/12/50, 23/5/51.
41. NRO C/ED 138, Logbook 2/4/51.
42. NRO C/ED 138, Logbook 8/10/51, 2/11/51.
43. NRO C/ED 138, Logbook 31/12/51.
44. NRO C/ED 138, Logbook: 19/9/52, 16/7/58.
45. NRO C/ED 138, Logbook April 1952
46. NRO C/ED 138, Logbook 3/10/51.
47. NRO C/ED 138, Logbook 21/08/50, 17/10/50, April 1952, 14/10/55, 26/5/59.
48. NRO C/ED 138, Logbook 27/10/55.
49. NRO C/ED 138, Logbook 16/11/55.
50. NRO C/ED 138, Logbook 16/11/55, 12/4/56.
51. NRO C/ED 138, Logbook 4/3/57.
52. http://www.historic-uk.com/CultureUK/The-1950s-Housewife/.
53. NRO C/ED 138, Logbook 21/3/57.
54. NRO C/ED 138, Logbook 12/10/50.
55. NRO C/ED 138, Logbook 28/3/60.
56. NRO C/ED 138, Logbook, 34/4/52.
57. NRO C/ED 138, Logbook 14/3/56, 26/3/56, 10/4/56, 24/3/57.
58. NRO C/ED 138, Logbook 28/7/51, 28/6/55, 7/758.
59. NRO C/ED 138, Logbook 15/6/56, 6/2/57, 6/3/58, 11/3/58.
60. NRO C/ED 138, Logbook 3/10/58.
61. NRO C/ED 138, Logbook 24/4/53, 7/2/56.
62. NRO C/ED 138, Managers Minutes 12/12/57.
63. NRO C/ED 138, Logbook 19/3/52, 13/4/55.
64. NRO C/ED 138, Logbook 10/1/49.
65. NRO C/ED 138, Logbook 15/6/50.
66. http://www.historic-uk.com/CultureUK/The-1950s-Housewife/.
67. Colloni, S. The Literary Critic and the Village Labourer: 'Culture' in Twentieth-Century Britain. *Transactions of the Royal Historical Society*, 2004, Vol. 14, p.107, Cambridge University Press, www.jstor.org/stable/3679308.
68. Whitbread 1972, p.109.
69. NRO C/ED 138, Logbook 23/04/56, 18/12/58.
70. NRO C/ED 138, Logbook 2/10/50, 29/07/59.

71. NRO C/ED 138, Logbook 18/07/58.
72. NRO C/ED 138, Logbook 2/05/50, 4/11/54.
73. NRO C/ED 138, Logbook 12/11/52, 9/07/54,16/01/56.
74. NRO C/ED 138, Logbook 1/09/5 5.
75. NRO C/ED 138, Logbook 17/07/5 7.
76. NRO C/ED 138, Logbook 5/09/55, 3/09/55, 16/11/56, 10/10/57, 13/03/5 8.
77. NRO C/ED138, Logbook 2/12/53, 24/06/54, 23/02/55, 7/05/56, 30/08/56, 8/01/57, 11/11/57.
78. http://www.historic-uk.com/CultureUK/The-1950s-Housewife/.
79. NRO C/ED 138, Logbook 26/02/59.
80. NRO C/ED 138, Logbook 30/01/58.
81. Sapwell, J. *A History of Aylsham*. Rigby, Norwich, 1960, p.48.
82. NRO C/ED 138, Logbook 19/02/59.
83. NRO C/ED 138, Logbook 12/03/56, 16/04/56).
84. NRO C/ED 138, 16/1/58.
85. NRO C/ED 138, Logbook 11/05/55, 11/10/56, 25/10/58.
86. NRO C/ED 138, Logbook 7/5/51, 11/02/57, 15/10 57.
87. NRO C/ED 138, Logbook 9/07/54.
88. http://www.historytoday.com/roland-quinault/britain-1950#sthash.KXerks41.dpuf Quinault, R., Britain 1950, *History Today* Volume 51 Issue 4 April 2001.
89. http://www.historic-uk.com/CultureUK/The-1950s-Housewife/.
90. Quinault, R., 2001.
91. NRO C/ED 138, Logbook 12/01/56.
92. NRO C/ED 138, Logbook 30/11/54.
93. NRO C/ED 138, Logbook 2/05/50, 25/09/51, 15/01/52, 24/11/52, 15/10/55.
94. NRO C/ED 138, Logbook 3/03/52, 15/07/55.
95. NRO C/ED 138, Logbook 5/3/51, 22/2/57.
96. NRO C/ED 138, Logbook 9/5/56).
97. NRO C/ED 138, Logbook 26/4/52.
98. NRO C/ED 138, Logbook 25/9/50, 16/4/51.
99. NRO C/ED 138, Logbook 31/8/59.
100. NRO C/ED 138, Logbook 17/05/49).
101. NRO C/ED 138, Admissions Book.
102. *Eastern Daily Press*, 25/2/58.
103. NRO C/ED 138, Logbook 6/7/59.
104. NRO C/ED 138, Logbook 11/3/57.
105. https://www.bankofengland.co.uk/-/media/boe/files/quarterly-bulletin/1994/inflation-over-300-years.pdf.
106. NRO C/ED 138, Logbook 2/3/57.
107. NRO C/ED 138, Managers' Minutes 6/5/57.
108. NRO C/ED 138, Logbook 28/4/52.
109. NRO C/ED 138, Managers' Minutes 12/12/57.
110. NRO C/ED 138, Logbook 26/6/51.
111. NRO C/ED 138, Logbook 22/7/51.
112. NRO C/ED 138, Logbook 23/7/51.
113. NRO C/ED 138, Logbook 27/8/56.
114. NRO C/ED 138, Logbook 12/11/57.
115. NRO C/ED 138, Logbook 12/11/57.
116. NRO C/ED 138, Log Book 25/11/54.
117. NRO C/ED 138, Log Book 25/11/54, 8/12/54.

118. http://www.nationalarchives.gov.uk/education/resources/fifties-britain/?show=all#introduction.
119. NRO C/ED 138, Logbook 29/5/52.
120. NRO C/ED 138, Logbook 23/4/58.
121. NRO C/ED 138, Logbook 18/2/49, 30/6/58, 6/5/49, 3/7/58.
122. NRO C/ED 138, Logbook 2/9/56).
123. NRO C/ED 138, Logbook Oct-Nov 57.
124. NRO C/ED 138, Logbook 26/6/50.
125. NRO C/ED 138, Logbook 28/4/49.
126. NRO C/ED 138, Managers Minutes 6/12/54.
127. NRO C/ED 138, Logbook 2/9/54.
128. NRO C/ED 138, Logbook 7/1/56, 29/9/58.
129. NRO C/ED 138, Logbook 8/2/51, 8/5/51, 29/6/51, 9/10/53. 7/4/54, 21/7/11, 28/11/55, 10/12/58.
130. NRO C/ED 138, Logbook 8/1/57.
131. *Eastern Daily Press*, 3/2/55.
132. NRO C/ED 138, Logbook 16/3/56.
133. NRO C/ED 138, Logbook 18/6/56, 26/10/58, 5/3/59.
134. NRO C/ED 138, Logbook 31/3/50, 4/4/50, 5/3/51, 16/3/51, 27/4/51, 4/5/51, 5/10/53, 16/2/54, 26/10/56.
135. NRO C/ED 138, Logbook 13/10/52, 23/2/53, 30/11/55, 18/6/56, 12/12/56, 4/1/60, 23/9/60.
136. NRO C/ED 138, Managers' Minutes 6/12/54.
137. NRO C/ED 138, Managers' Minutes 6/12/54.
138. NRO C/ED 138, Logbook 23/9/53, 21/3/58, 5/5/59.
139. NRO C/ED 138, Logbook 14/1/52.
140. NRO C/ED 138, Logbook 14/11/52.
141. *North Norfolk News*, 20/10/50.
142. NRO C/ED 138, Logbook 8/11/57.
143. NRO C/ED 138, Logbook 7/12/55.
144. NRO C/ED 138, Logbook 14/3/56.
145. NRO C/ED 138, Logbook 29/6/51, 15/1/57.
146. NRO C/ED 138, Logbook 29/10/57, 19/6/58, 14/12/59.
147. NRO C/ED 138, Logbook 3/11/55, 1/11/56, 8/10/58, 3/3/60.
148. NRO C/ED 138, Logbook 10/10/55.
149. NRO C/ED 138, Logbook 18/5/55.
150. NRO C/ED 138, Logbook 3/12/52.
151. NRO C/ED 138, Logbook 14/11/60.
152. NRO C/ED 138, Logbook 13/9/54, 12/3/53.
153. NRO C/ED 138, Logbook 4/5/55.
154. NRO C/ED 138, Logbook 7/12/5.
155. NRO C/ED 138, Logbook 25/9/60.
156. NRO C/ED 138, Logbook 27/10/60.

Primary Sources
Norfolk Record Office (NRO)
NRO, C/ED138, Aylsham Community Nursery School 2001:
    Box 1: Log Books 3 Jan 1949 – 13 Oct 1955; 14 Oct 1955 – 5 Dec 1960.
       Head teacher's Reports to Managers 18 Mar 1954-18 Jul 1960 (listed as Managers' Minutes).

Box 2: Pupil Admission Records.
NRO, C/ED 81/5, Rules of Management.
NRO, MF747:23-32,Aylsham Tithe Map 1839.
NRO, MF1559/3 St Georges Infant School, Great Yarmouth.

Norwich Heritage Centre
North Norfolk News, 20 Oct 1950.
Eastern Daily Press, 3 Feb 1955.

Secondary Sources
Aylsham Local History Society, *A Backward Glance*. ALHS, 1995.
Boyce, E.R. P*lay in the infant school, Number Rhymes and Finger Games*. Methuen, London, 1949.
Palmer, A. Nursery Schools for the few or the many? Childhood, education and the State in mid-twentieth century England. *Paedagogica Historica*, 2011, Vol. 1-2, pp.139-54 (off print).
Sapwell, J. *A History of Aylsham*. Rigby, Norwich, 1960.
Whitbread, N. *The Evolution of the Nursery-Infant School*. Routledge Kegan and Paul, 1972.

Digital Sources
Legislation
http://www.legislation.gov.uk/ukpga/Geo5/8-9/39/contents (Education Act 1918) accessed 2014-2017
http://www.legislation.gov.uk/ukpga/Geo6/7-8/31/contents (Education Act 1944) accessed 2014-2017
http://hansard.millbanksystems.com/written_answers/1943/jul/22/war-time-nursery-wardens-educational accessed 2014. (*Hansard*, HC Deb 22 July 1943 vol. 391 c1093W)
http://hansard.millbanksystems.com/commons/1945/mar/09/war-time-nurseries accessed 2014. (*Hansard*, HC Deb 09 March 1945 vol. 408 cc2425-50)
http://www.historic-uk.com/CultureUK/The-1950s-Housewife/ (Castelow, E., 1950s Housewife) accessed 2014-2017
https://www.bankofengland.co.uk/-/media/boe/files/quarterly-bulletin/1994/inflation-over-300-years.pdf accessed 2014-2017 (MacFarlane, H, and Mortimer-Lee, P. Inflation over 300 years, *Bank of England Quarterly Bulletin*, May 1994, pp.156-162).
www.jstor.org/stable/3679308 accessed 2014-2017 (Coloni, S., The Literary Critic and the Village Labourer: 'Culture' in Twentieth-Century Britain. *Transactions of the Royal Historical Society*, 2004, Vol. 14 (2004), pp. 93-116, Cambridge University Press).
http://robert-owen-museum.org.uk/New_Lanark_Schools accessed 2014-2017
http://www.historytoday.com/roland-quinault/britain-1950#sthash.KXerks41.dpuf accessed 2014-2017
http://www.nationalarchives.gov.uk/education/resources/fifties-britain/?show=all#introduction accessed 2014-2017
http://www.norfolkmills.co.uk/aylsham-navigation.html accessed 2014-2017

Biographical note

A founder member of FACHRS, Stella Evans was on the Management Committee until 2006. She coordinated both the Swing and Emigration projects and she has participated in the research for other FACHRS projects. DA301 was the final course that she studied before graduating from the Open University with a BA(Hons). She went on to obtain her taught Master's degree in Local and Regional History from the Centre of East Anglian Studies at UEA.